The Origin of Specious Nonsense

by

John J May

ORIGINAL WRITING
www.darwinsdeadidea.com

© 2010 John J May

All rights reserved. No part of this publication may be reproduced in any form or by any means—graphic, electronic or mechanical, including photocopying, recording, taping or information storage and retrieval systems—without the prior written permission of the author.

978-1-907179-71-6

A CIP catalogue for this book is available from the National Library.

Published by Original Writing Ltd., Dublin, 2010.

Printed by Cahill Printers Limited, Dublin.

DEDICATIONS

I dedicate this book to my ex-wife Yvonne who taught me the meaning of tolerance and unselfish love,

And

Our six beautiful children who showed me the pleasure of simplicity and joy.

And

Our four magnificent grandchildren who are teaching me appreciation and the daily beauty of life in all its glorious expressions.

"The most formidable weapon against errors of every kind is reason."

THOMAS PAINE (1737- 1809)

Reason is betrayed by what eminent writers-scientists-philosophers and thinkers wrote about Darwinian evolution; "A Hoax"- "Joke" - "The greatest deceit in the history of science." - "A fantasy" - "An impossibility" - "A fiction" - "A fairytale for adults."

You will be shocked, mocked, amazed, dazed, confused, amused, enraged, engaged...but most of all thrilled and mentally fulfilled by the information you are about to read.

The deadly daily drip of pernicious poisonious percolation of Darwinian evolution contaminates simplicity and pollutes reason. It is time for common sense and the known evidence of science to energise the millions of educated people in this world to stand up and say;

ENOUGH OF THIS SPECIOUS NONSENSE.

CONTENTS

Introduction   ix

1 SANTA CLAUS   1

2 ORGANISED RELIGIONS   5

3 IMMORTALITY OF THE SOUL   13

4 HELLFIRE   18

5 MAGIC AND PSYCHICS   25

6 HOLY BOOKS DICTATED BY GOD   31

7 EVOLUTION   38

Introduction

To undertake the extirpation of fond fictions from the mind is, I know, irrefragably fraught with explosive consequences. Therefore, I begin as I mean to finish, gently, with simple explanations for complex concepts and hopefully to elevate reason and true science as a magnate to sanity, purpose and a future with hope.

All human beings suffer. Not to suffer is not to live. Worriers endure more because they imagine more. One of the softest organs hidden in our flesh, the heart, suffers hardest because apparently it is capable of emotional feelings!

A shattering example of life's trial-some tragedies is the abandonment by a mate once loved. It brands the brain with desperation-confusion-fear-accentuated loss-loneliness and many other inexplicable aches. After time the heart erects a frightened fence to protect against repeat performances.

There are innumerable other casual sudden shocking experiences which the innocent suffer and endure during our average 25,550 days living on this amazing planet which to enumerate is unnecessary: suffice to quote the famous dictum: "Most men lead lives of quiet desperation." It is with this in mind that my book is written conscious of the unnecessary pain added by imaginary fictions to mans daily struggle for meaning and some happiness.

In the beginning (No one truly knows when that moment in time began!) (God -The COGNITIVE ARTISTIC GENETIC ENGINEER) created the universe and our spaceship called earth! Since then (then, being when man and woman first appeared) we have had chaos-murder-wars-confusion-delusion-pain-suffering-destruction-heartache-and a relatively short lifespan.

There are certain questions human beings muse over to one degree or another, at some point during life, and particularly so when dying.

7 QUESTIONS:

How and when did the universe originate?

Is there a God?

Why am I on this planet?

What is my purpose here?

Why must I die?

Where do I go when I die?

Why is there so much injustice and pain?

These questions have baffled the intellect of philosophers, theologians, scientists and ordinary mortals for centuries.

Every person throughout history was born without language-hate-beliefs-phobias-bias-religion-pretension-tribal or political loyalties-pre-conceived ideas-prejudice et cetera and is predisposed to believe ANYTHING, taught by parents, teachers or local community. . . .

We accept things whether logical or nonsensical through language and environment. For example the Moslem noble teaching that all men are brothers going back in time to one man and woman is beautiful when elucidated by various Imams, and to its credit, it must be said that the Moslem faith is virtually free from racial prejudice.

However, the reality of equality-fairness-brotherliness and mercy disappears from Moslem hearts when political differences erupt as in Iraq and other places where Sunni's, Sufi's and Shi-ite's vie for political and religious dominance! The murderous merciless mayhem each side inflicts is a reflection of childhood influences. A religious slogan said: "Give us a child and by the time he's seven, he shall believe he's on his way to heaven!"

Another unscientific and illogical belief which no educated adult on first hearing would or could embrace—even though it is ostensibly believed by one billion people on Sunday mornings throughout the world. Men dressed in women's black clothes in Roman Catholic Churches, lighting candles, ringing bells and mumbling incantations: Then "magically" the chemical composition of flour/bread is said to become human flesh and entire countries almost, in some places, willingly engage in what is universally condemned as revolting and illegal: cannibalism! It is true that this act of communing with their (whether they like it or not) Jewish leader, Jesus, makes them feel 'holy' and sometimes does produce feelings of felicity towards others!

This 'sense' of magic and unity does not make it true . . . In fact dying for a belief won't make it genuine if it is false! The question is how can Muslims sincerely claim to believe in brotherhood with their lips yet their hands slaughter each others families? How can Catholics not admit that bread before, during and after the liturgical service is, was and always shall surely remain bread and not the flesh of a brilliant young Jewish Rabbi who was mercilessly murdered 2000 years ago! The uncomfortable shuffling silences when questioned on these fantasies are an eloquent testament to the known realities. Those in both systems of theological treacheries who are thinkers know in their hearts, 'something is wrong!' Those who do not think are condemned to obligatorily pass those fallacies seeping silently into their children's psyches through ossified osmosis onto the next unfortunate generation . . .

The dictionary definition of fiction is
"SOMETHING INVENTED BY THE IMAGINATION."

Possibly excepting Santa Claus those invented seven fictions have caused heartache and suffering for millions. This book shall realistically and scientifically demolish destructive beliefs and replace them with understandable constructs which unite hope with reason, purpose with passion, logic with science, facts with reality and knowledge with understanding.

THE ORIGIN OF SPECIOUS NONSENSE

The Scream by Edvard Munch

Every single tradition, idea, belief or myth has an origin in the past. *Nothing* no matter how 'sacred' or revered should be exempt from forensic analysis and methodical inspection. Things are not what they appear and what seems like nothing might contain the seed of an indisputable unassailable truth . . . Why should we be afraid of what is demonstrably true? Or fear the origins of what we might hold dear, be it religion, science, traditions, political persuasions or evolution? Part of loosing our grip on cherished ideas is, pride, loosing face, (whatever that means!) confusion, peer pressure and a myriad other influences only we are privy to! A community of co-believers is a powerful magnet for the soul. It exists in religions, political parties, scientific fraternities, golf clubs and thousands of other groups where to one degree or another atmosphere of 'us and them' is fostered and becomes ultimately the cement that bonds. . . .

Questioning in moderate groups is frowned upon and in those who "know they possess the truth" punished by excommunication or even death! It takes courage to be different, to want to know truth, facts, to seek out answers, the whys and how's, when and where's of origins, it may not endear us to ones family or peers but it certainly makes one more at peace with the inner self. ("Seek the one who seeks the truth—but ignore those who claim they found it.")

This book shall attempt to answer seven eternal questions without guiding anyone to any church, group or religious organisation. The only one we should 'belong' to is oneself and the only intellectual loyalty we should give is to scientifically demonstrable truths! Metaphysical speculations are meaningless to the starving, pointless to the confused, worrying to the innocent and of no real interest to the groaning masses. As Ludwig Wittgenstein once remarked, (and he may well have had Immanuel Kant in mind), "that reading philosophy is A KIND OF AGONY." He was a brilliant philosopher who delved into the meaning of language, words and feelings. Very much an individualist he had moral courage to publically proclaim his belief in a future life.

The impression no intelligent and modern person could possibly believe in this concept is so firmly rooted it is worth saying this is not the case.

Wittgenstein's philosophy provides little support for traditional Catholicism, he nevertheless wrote:

"What inclines me to believe in Christ's resurrection? It is as though I play with the thought. If he did not rise from the dead, then he decomposed in the grave like any other man. He is dead and decomposed. In that case he is a teacher like any other, and can no longer help; and once more we are orphaned and alone and must content ourselves with wisdom and speculation. We are, as it were, in hell, were we can only dream, and are, as it were, cut off by heaven by a roof." Quote from Priest Jonathan Robinson, in the 'Mass and Modernity.'

One can think what one likes about Wittgenstein's argument; my point here, though, is that he was surely an intelligent person, and did not find the notion of Christ's resurrection to be inherently absurd or unthinkable.

Ultimately we all stand alone and occasionally feel emptiness depending on background and the number of emotional blows we have taken! Sometimes for whatever reason we might wonder, 'What is the point of it all?' Most times the blackness lifts but occasionally and catastrophically it doesn't! Conflict appears to be part of our human psyche and an analysis of our past confirms it. People with common sense (which is not that common) decide early in life to don a mask and play a game of social interaction to avoid as much as possible arguments and their often dire consequences . . . Life is short, often brutal and for the majority (even in the west) a daily struggle to pay bills and hope to find some meaning and a little love. This age of information and choices is still riddled with pain and confusion, regardless of comfort and wealth enjoyed by half the world's population, one reason being the moral compass is confusing.

Wittgensteins profound concept is based on a simple premise. If there is a God there is a future existance—if no God, no future life.

I
SANTA CLAUS

Millions of innocent children excitedly go to bed each year on December 24th believing a jovial white bearded old man dressed in red will leave them a gift as they sleep. The logistics escape them and the origin of this touching story is an irrelevancy! He comes, he goes, he leaves presents and next morning their irrefutable faith has been rewarded as parents and guardians worldwide collaborate in this apparently harmless hoax! It is one of Christianity's time honoured fables which genuinely brings happiness and joy to millions of delighted children and parents . . . In the minds of those young human beings his existence is an "indisputable fact!"

The origin of Santa Claus is connected (by tradition) to a man who acted as a "Bishop" of Myra in present day Turkey. Apparently he died for what he believed, and human nature being what it is, he was called "a saint" by generations who lived after his alleged murder in 350 A.D. They, thereafter, concocted various stories about him.

However, those of us who have reached the age of reason know for a certainty based on empirical evidence and in spite of pleas for reassurance mingled with tears and bitter disappointment—*Santa Claus is a fiction.*

Thousands of years ago most people were ignorant, illiterate and fearful of the seasons. Over time the idea was formulated that the sun was in fact a god that needed to be appeased. This nonsense was articulated by uneducated priests who soon realised, people were mollified when they spoke "mumbo jumbo" with a mixture of observations threats and rewards.

December 25th during Roman times was a day of riotous celebration and debauchery because ignorant priests were able to convince masses of uneducated people that the sun, a major god, was not dying and would noticeably rise in the sky and soon shower down heat and life and growth. Constantine, the pagan emperor, was sickened by the mind-bogglingly, boring (but vicious among the clergy) nonsensical ongoing arguments about the nature of the Jewish man who had caused trouble in Israel 300 years ago! Baffled by this theological clap-trap he called a religious council in Nicaea in 325 A.D. and *he* decided Jesus was God and those who disagreed with his decision were in serious trouble. Oh, incidentally this Roman emperor who was steeped in pagan Greek philosophy, also for good measure decided which letters God had written to his new people now called Christians. So what were they going to do with the holly,—ivy,- mistletoe, -Yule logs, et cetera! Teach that the Jewish rebel Jesus was born on our festival day December 25th!! Problem solved . . . One thousand years later the Catholic Church had complete control over most of Europe and the various royal families all kow-towed to Rome.

In the seventeenth, eighteenth, nineteenth and twentieth centuries (especially after the publication of -"*A Christmas Carol*" by Charles Dickens) the traditions surrounding Christmas took on a life of their own with cards, gifts, drink, presents and a lovely little fiction—that Santa Claus brings gifts to millions of innocent little children on Christmas morning . . .

There is a reason why I chose this as the first of my seven fictions to demolish! It sets the stage for the next six. These will be more difficult to extirpate because religious roots go deeper, however the principles of origins, evidence, traditions, respect, customs and affection remain the same. Not one person living asked to be born and it is an axiom that if we had been born in a different country with different religious, cultural, political and social mo-res then that is the country we would adopt and those beliefs endemic to that locality would be ours. Make no mistake about the absolute validity of that sociological fact; which begs the question: are my beliefs my own or were they automatically (gently or otherwise) foisted on me?

Seven questions might also be asked!!

Am I racist?

Am I prejudiced?

Do I get angry if my beliefs are challenged?

Do I know the origins of my beliefs?

Am I afraid to question my faith?

Was I encouraged to think as a child?

Shall I be punished for questioning?

There are seven other questions directly connected to the above.

Am I happy?

Can I accept life as it is?

Have I a purpose?

Have I found love?

Have I squandered love?

Have I discovered what is truly important?

Do I have a future?

Fundamentally we are all the same, tragically we are superficially divided and those divisions need not seem so very important when we have origins revealed. We were born to live and love but so few truly do! Why? I have experienced spectacular happiness (even though from a poor background, as a child does not know what poor is if he/she is loved) and suffered monumental loss! However every day is a day to live, love, learn and appreciation of each one enhances its intrinsic priceless value . . .

Our childhood may have been idyllic or painful, however the simple fact remains: *Santa Claus is a fiction!*

2

ORGANISED RELIGIONS

Organising people to express reverence, unity, awe, admiration and praise for a god or gods in an atmosphere of subservient acquiescence has caused chaos,- fear.-hatred,—bloodshed and bitterness since Cain murdered his younger brother as recorded in that uniting/divisive (an oxymoron, I know!) book known as the Bible. Throughout history the cruellest savageries have been inflicted on "unbelievers" and families have been and are torn asunder by theological treacherous tyrannies! This toxic poison is made more potent by ancient writings by charlatans'—desert savages—uneducated, unedifying, unfulfilled delusional delirious maniacs! Others who wrote some of these works were, admittedly, sincere educated men and women who had the good of people at heart. Reading through much of this drivel is laughable and yet saddening. Sometimes some parts of those books appear to ignite a desire to change and help others, and also explain mans' origins.

I am a recovering Roman Catholic and had the misfortune as a young child to be infected by dislike for other religious persuasions—, Protestants in particular, and terrorised by the hellfire doctrine of unimaginable cruelty, to last—yes—forever. (Which, if true, should make us, if we are honest, actually hate God.)

To fully appreciate the infectious calamitous pox organised religion has inflicted on billions of children read the brilliant book—'*The End of Faith*'- by gifted writer Sam Harris published in 2005: In particular chapter one, '*Reason in Exile*'—if ever there was medicine for the mind to diminish the fever of toxic childhood chilling chimera's, this is surely it.

Were you taught as a child that the Bible, Koran, Book of Mormon, et cetera, are literally words from God or 'Gods word'? If you were you might agree there are in those writings some sound moral principles that have guided your life for good!

The Jews after King Nebuchadnezzar's Babylonian captivity around, 600 B.C developed the tradition of meeting in homes, hovels and houses of 'worship' called Synagogues. History of their ancestors was here written—discussed—debated—analysed and argued over as recorded by others and interpreted, during the past 1000 years. The Jewish encyclopaedia edited by Cecil Roth and Geoffrey Wigoder published 1959 under 'Exile Babylonian' states; "The deportees in Mesopotamia mostly peasants and craftsmen, with their main centre at Tel Abib near the great canal, were encouraged by the confident counsels of Jeremiah and later inspired by the ecstatic prophecies of Ezekiel.

Hence, they were able to maintain and fortify their religious identity. The circumstances of Babylonian Exile seem to have given rise to the institution of Synagogue and, as a corollary, to the beginnings of the prayer-book and of the *canonization of the scriptures.*" "Teachers" and "Rabbis" became the anvil by which most ideas were hammered into organised belief and submission.

The same process could be said to be at the core of most organised religions. Once a few sincere hypocrites decide to 'meet for meat of the word' someone, usually male, will start to dictate what is acceptable and what is objectionable. . . . Oscar Wilde (Irish play-wright. 1854–1900) said: "Religion is the fashionable substitute for belief" or another of my favourite writers Christopher Hitchens (Author—athiest—controversialist.) wrote in his fascinating book "God is Not Great" page 6—7 under the heading 'Putting it Mildly' published in May 2007. "There is no need for us to gather everyday or every seven days or on any high and auspicious day, to proclaim our rectitude or to grovel and wallow in our unworthiness. We atheists do not require any priests, or any hierarchy above them, to police our doctrine. Sacrifices and ceremonies are abhorrent to us, as are relics and worship of any images or objects (even including objects in the form of one of man's most useful innovations: the bound book). To us no spot on earth is or could be "holier" than another to the ostentatious absurdity of the pilgrimage, or the plain horror of killing civilians in the name of some sacred wall or cave or shrine or rock, we can counter pose a leisurely or urgent walk from one side of the library or the gallery to another, or to lunch with an agreeable friend, in pursuit of truth or beauty. Some of these excursions to the bookshelf or the lunch or the gallery will obviously, if they are serious, bring us into contact with belief and believers, from the great devotional painters and composers to the works of Augustine—(Catholic 'Saint'), Aquinas (Catholic 'theologian'), Maimonides (Rabbi—writer—teacher, who catalogued the 613 Jewish laws and precepts), and Newman (Roman Catholic Cardinal). These mighty scholars may have written many evil things or many foolish things, and been laughably ignorant or the germ theory of disease or the place of the terrestrial globe in the solar system, let alone the universe, and this is the plain reason why there are no more of them today, and why there will be no more of them tomorrow. Religion spoke its last intelligible or noble or inspiring words a long time ago: either that or it mutated into admirable but nebulous humanism, as did, say, Dietrich Bonheoffer, a brave Lutheran Pastor hanged by the Nazis for his refusal to collude with them. We shall have no more prophets or sages from the ancient quarter, which is why the devotions of today are only the echoing repetitions of yesterday, sometimes ratcheted up to screaming point so as to ward off the terrible emptiness."

Organised religions destructive venom poisons everything and this toxic unreasonable parasite clamours for our crawling submission, subservience and devotion. I could quote from many books and some philosophers who saw the intrinsic moral corruption of, rewards for belief versus punishments for unbelief! However in the past 5500 years (approx) of *known* recorded history, archaeologist's have uncovered some of the oldest cuneiform writings impressed into stone tablets from Summeria, Akkadia, Assyria, Babylon, Persia and Egypt: these usually are commercial transactions, records of kings, battles and in particular mentions of 'an after life' in The Egyptian 'Book of the Dead' being a good example of invented fictions palmed off on the frightened dying as fact because Isis the Goddess of fertility had spoken! (Or any one of dozens of other gods to choose from).

Regardless of what we believe about Holy Books like the Bible or Koran et cetera, many things written in them by men are simply helpful, revealing and inspirational! The question needs to be asked: where did the idea originate that gathering together in a building for collective repetitive liturgical mumblings is something asked of us by our respective god! (or gods)! If he is the supreme scientist—architect—mathematician or—designer of our staggeringly complex and immeasurable universe, why do we diminish his grandeur by imagining he can be contacted in a large stone or wooden building?

I want my six children to be happy. I do not want them to gather together on a fixed day of each week or month or year, sanctify the building they gather in, (in my name) and start praising me . . . that would be embarrassing. No, what I prefer is that they show love to their mother, kindness and affection towards one another and whenever my name comes up if they feel so inclined to express genuine appreciation for what I willingly did for them as they grew to responsible adulthood. All I ask as a loving Father (to which the six testify on occasion) is from their own volition express love and appreciation towards me when and if they feel like it . . . that is my reward, that is all I ask. . . . .

I cannot conceive of a super intelligent, amazingly loving God or Father (Father meaning "life giver") wanting his creation to be mired in ignorance, steeped in superstition, sunk in confusion, drowning in despair, which for the most part is true in religions. Ignorance is expensive and knowledge relatively cheap. Theologians and philosophers have struggled with the question of Gods existence and nature for centuries. . . . . The majority of people hold sincere, valid opinions of which many are, no doubt, accurate on various topics. However, on the subject of organised religion the one thousand or so competing Christian sects alone would hold, often, violently divergent views on how 'worship' should be conducted. I speak of Christianity here since I was born into Roman Catholicism, and converted into another with a pathological aversion to blood and rancid apoplectic hatred for Christianity. (Especially those who once were members and had the audacity to ask normal questions of abnormal theological constructs) I refer to Jehovah's Witnesses as founded by an end-of- the-world interpretation freak, Charles Taze Russell, who if alive today would be horrified at the way the Witnesses have institutionalised cruelty for ex members who had the courage to sever connections with men who poison the simple teachings of Jesus. This Rabbi is recorded in a book as saying, "I will never turn anyone away, anyone who comes to me." Did he actually say it? I don't know for certain! First of all I need to establish if he actually lived two thousand years ago. After I establish the authenticity of his existence, can I for certain prove the veracity of the above saying? Finally if I am attracted to this unusual Jewish Rabbi's powerful teachings of love, kindness, forgiveness and hope can I just live the rest of my life in peace trying to help my fellow man? After all he is also recorded as having said, (Note: not commanded) "For where two or three come together in my name, I am there with them." So why do I have to join a church—group—congregation? For example, if I meet in a pub or bar with one or two agreeable friends and drink some wine or beer, and we discuss our undying admiration for this Jewish rocking rebel, and how his moral teachings encourage us to be loyal in love, kind in conflict, forgiving, tender with children, helpful to the poor, humble towards humans, compassionate to animals and in particular realistic in our realities. Surely, this is the true spirit of his message.

    Maimonides counted 613 Jewish legal precepts known as 'THE LAW OF MOSES.' When Jesus was asked what was out of all these the greatest command? (Or Law)—His answer was, 'To love God with all our hearts (easy), soul (puzzling), mind (difficult) and strength (how?). He continued: 'To love your neighbour as yourself.' That is almost impossible . . . hold on Jesus . . . that's heavy . . . I don't even like some of my neighbours. In fact I don't know most of them! However who knows anyone!

The Four YOUS

There is the you, you know!

The you, you don't know!

The you, you want to know!

And the you, you don't want anyone else to know!

First I must love myself! I don't think traditional Christianity likes that metaphysical concept! Nevertheless, I know, I do—so now I must equate that great feeling into a moral context and 'love my neighbour.' But lets get real here and ask who realistically loves (yes loves) their neighbour the way we love ourselves? I certainly do not and cannot. A relative called yesterday and told me he needed a place to live! I heard he is a liar and a thief but he was still in need. Did I offer him a spare room in my house (which I had). No, I hypocritically took his phone number and gave no real help. Does that make me a poor follower and bad example as an admirer of Jesus? Most assuredly yes!!

I read a book recently published called '*The Purpose Driven Life*' by a good American named Rick Warren. On the cover it states "The Bestselling Non-fiction Hardback Book in History" (30 million sold worldwide). What I liked about this man is he is helping to feed the hungry in dozens of countries.

His history is suspect. His theology born from childhood. His conclusions traditional. His background springs from the greatest hatred towards black people in history (sanctified by texts from his mandate, the Bible, of course) However, he shows love and appears to mean it!

Time magazine in the summer of 2008 featured him on its cover and ran an interesting story. I sent the following letter which they did not publish!

A MAN FOR ALL SEASONS:

"As a visitor to the U.S.A. I was impressed by Rick Warren's take on Christianity (18/08/2008 *Time*). As a European I highly value the lowly esteem by which we view 'Religious Leaders.' However, here is a man who appears instinctively to understand the powerful injunctions of a poverty stricken rabbi murdered 2000 years ago! If he accomplishes half of what he is trying to do he will unleash the dormant latent love in millions of human hearts for the less fortunate on this crying dying sighing planet.

He has opened my closed mind to the magnificent magnanimous majesty, committed Christians might accomplish when organised to, "do unto others as you wish them do to you."

Maybe this man of unshakable purpose might on the world stage help remove some of the historic blame and shame which clergymen have caused to that Jewish man's name, and descendants.

I for one applaud your noble efforts, admire your innocent dreams and am astonished by your courage."

N.B: Power Corrupts

John J. May
Dublin, Ireland

Jesus never read the Greek scrolls known as The New Testament. Abraham never read Hebrew scrolls called, The Old Testament. The first male and female never read either of them! So it appears that meeting on a regular basis to read and hear words written thousands of years ago actually hinders human harmony, hardens hearts, and poisons peace. At best, it gives hope and at worst causes war.

If I impertinently ask someone how many times a week do they passionately kiss their partner? I should rightly be told to mind my own business, as such a gesture between two people is precious and private! So it should be with any worship we give. Private . . .

If the human race are brothers and sisters, related through blood and dreams, then surely, 'peace towards all men' should be our passion!

3
Immortality of the Soul

Atheists have no problem rejecting this ancient pagan fictitious notion of the soul's immortality . . .

Before any so called "Holy Books" were penned this understandable fiction was believed, "when you die you do not die!" Why? Because life is precious, so priceless in fact that people have literally killed so they could live. A child usually has no concept of death! The average person seldom contemplates old age or dying, reason being we were designed on a biochemical level to exist as blood and flesh forever. If the cellular/chemical composition of a young man or woman could be maintained then molecular breakdown, i.e. wrinkles, et cetera, *would not occur*. So, the raging desire to prevent aging is genetically hardwired into our psyche, death is not natural. Animals, birds, fish and flowers, et cetera all die. True. But we are not of those species. It is because of our love of family, youth, friend's, health, music, food, knowledge, travel, discovery, chocolate, theatre, poetry, books, science, pleasure, conservation, comfort, alcohol, sex and all the other delights of living that we dread, fear, panic, and have a horror and terror of dying. The reason is simple—we know in our hearts, the dead are not living.

There are seven reasons why the Dead are Dead...

The body ceases breathing.

Flesh is ice cold.

Heart stops beating.

Remains rot in clay.

Clay contains 113 chemical elements (approx).

Cremation of a human produces 113 chemical elements.

We know the Dead are dead and science confirms it chemically.

There are seven reasons why people believe the dead are alive.

They want to.

They cannot imagine death being the final end.

They are unaware of ancient Egyptian priests inventing the calming notion of an unseen part (soul) departing bodies at death.

Immortality of the soul is attractive.

Religion teaches it.

There "must be something else!"

Living 'somehow' after dying is appealing.

JOHN J MAY

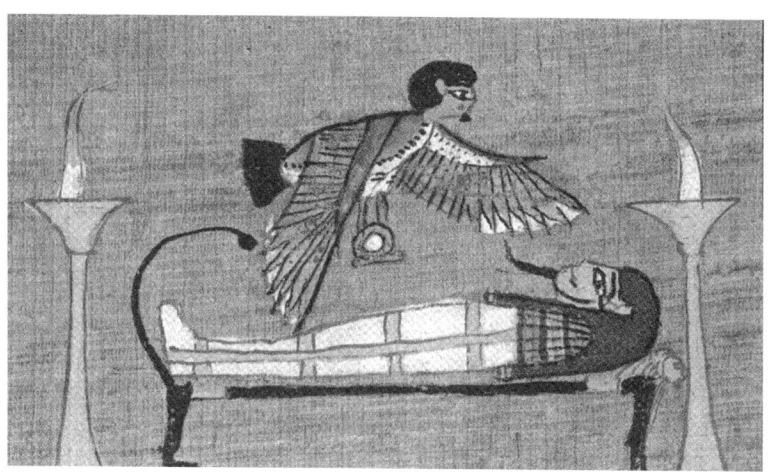

An artist–priest's idea of "something" leaving the body after death.
From: *Great Ages of Man: Ancient Egypt*
Published by Time—Life Books 1972

Underground Animal/Human gods judging the dead
From: *Ancient Egyptian Book of the Dead*,
translated by Raymond O. Faulkner
Published by Barnes & Nobel—New York (2005)

The scales in this ridiculous picture is the Egyptian belief that when death arrived the human heart was weighed. If the dead person got a bad result, he was condemned and a good one meant a life of bliss in the underworld. These silly ideas were reinforced by the Pharaohs and promulgated by the bald ignorant priests for prestige and profit. To the poor average Egyptian those fictions struck terror into his heart.

It is only when we stop and think about origins of concepts, ideas, views, traditions that we realise that notions such as 'Immortality of the Soul' begin to flounder. The American Heritage College Dictionary definition of soul is, "A human being" also, "The animating and vital principle in human beings, credited with the faculties of thought, action and emotion."

Seven Questions

Where did the soul come from?

Where does it go?

If at the exact moment of conception—it is tiny, so does it grow with the embryo?

If a woman miscarries what happens to it?

If a baby is born sick is the soul sick?

Is the soul (if immortal) a sentient thinking being?

Who invented this idea?

Let us now face reality and accept the following scientific facts. We no more possess an immortal soul than we do a guardian angel (If true they could be accused of dereliction of duty when bombs go off in Iraq and elsewhere.) No one reading this can give valid proof of an immortal soul which was alive 1,000 years ago and is still alive. In 1000 years time not one person who is living now will be able to prove unambiguously that a part of him/her shall be alive then.

We are made from birth of flesh, blood, bone and to survive need oxygen. When we stop breathing we cease to be. All the so called holy books in the world make no difference. In fact the Jewish book Ecclesiastes (allegedly written by King Solomon) scientifically states: "The living know they shall surely die, but the dead know nothing. . . ." The dictionary definition of nothing is, "Something that has no existence."

The science as recorded in the Jewish document Genesis is astounding: The fact that foolish fundamentalists of the Christian variety, in particular, embarrass us by their literal interpretations of the age of the universe and earth is of no consequence. No one truly knows for sure. And personally I don't care. The energy and passion wasted on superfluous theological arguments is like two dogs fighting over a bone. This document states that man was constructed from 113 chemical elements from clay and we at this moment in history return and turn to clay at death. However this record of mans origins state that when man was formed by the COGNITIVE ARTISTIC GENETIC ENGINEER; GOD *"Breathed into his nostrils the breath of life and the man became a living soul."* That means when that fool died he was a dead soul. Will that man live again at some point in the future? I don't know and selfishly don't really care. That decision rests entirely with God. What I do know for sure as a positive scientific fact is that the idea of humans possessing a mysterious everlasting item/thing inside their flesh called by religions mostly, an immortal soul, is an immoral *FICTION* . . .

4
HELLFIRE

This fictional idea has poisoned innocent hearts for thousands of years and terrified people with a morbid dread of the Creator. Imagine living in a culture with virtually no education, no scientific methodologies of evaluation, no concept of a purposeful existence! A short brutal dirty life dominated by brief moments of pleasure in daily dreary drudgery! An existence filled with fear of displeasing your unchosen God or gods!

Thousands of years ago most human beings were chained to the soil. Gods of seasons' growth and harvest's were dreamed up and worshiped in various invented ways. Also deities of the womb/fertility developed and people would offer some "first fruits" to their non-existent fictitious gods in order to appease them! Children listened to the fear of punishing fire as spoken sincerely by their deluded parents, grandparents, other children and of course ignorant preachers—priests and imams. Without hospitals, hygiene or modern methods of miraculous medication, burns were one of the most excruciating pains any human could suffer. And there it was—the terror of terrors to terrify the terribly ignorant into unquestioning obedience to one's cultural and religious raging, ranting precepts—as espoused by untutored clerics.

Ancient pagans in many cities used fire one way or another as symbols of purity, power, destruction, fear and sacrifice. For example the Canaanites and peoples under their influence sacrificed human beings, especially first born children (like me) to a fictitious god called Moloch by passing them through fire. (Imagine!) For most of recorded history people did not travel far beyond their huts/homes or caves as there were few modes of transportation as we understand it today. Any rubbish accumulated would have likely been discarded stored burned or buried.

However, large towns needed some form of disposal for waste and rubbish and southwest of ancient Jerusalem there was a valley called Gehinnom (also known as Gehenna) where the town's refuse, dead dogs, suicides and dead criminals were thrown into a fire which was constantly burning. This was a hygienic way of disposing of the waste of a city which is constantly made by humans and animals. About 35 years after the murder of Jesus by the Roman authorities in cahoots with Jewish religious and political leaders, some of his followers mentioned this fire/dump outside their then- functioning city in passing, and lo and behold the sick, sad, savage, shit idea of an everlasting hellfire took off like forest fire in a drought. The splits and malice in "Jewish / Christianity" were already evident and what better "holy terror" to frighten opponents with than fire. No pain was too great for apostates. The leaders of the various sects of the young Jewish rabbi knew they were to act like innocent little children, doves of peace and lost lambs! However as Mohatmas Gandhi (Indian political non violent leader) said almost two thousand years later, "Christianity would be nice if it was tried." If Hell (Fire) was true then no hatred, cruelty or malice would be enough for those unfortunates viewed as "Gods enemies."

Seven Hellish Fruits:

HELL the seed that feeds the greed of religious power.

HELL the fuel that fed the tool of torture.

HELL the "good" that fed the blood of violence.

HELL the fuel that turned mens' hearts so cruel.

HELL the chalice filled with pious sacred malice.

HELL the whip which tightened clergy's grip.

HELL the brew that punished the suffering Jew.

There is no more cruel wickedness than a parent who for whatever reason burns their child! Ancient priests did it! Ancient writings record it! The Nazi's practiced it and the Protestant and Catholic clergy were mostly silent during that sick savage sad reign of terror for twelve agonising years of tears and fears, of fiery militaristic hell on earth.

There is no HELLFIRE on the Moon, Mars, or under the soil we walk on . . . It is an assault of the good name and character of YHWH. It is an outrage, an affront, an insult, a vindictive fiction invented by ignorant men to terrify innocent people.

They quote their various holy books as evidence of this pox and ignore the soft gentle voice of their hearts and reason that plainly whispers "No loving God/Father would do this to his innocent children." But they make this God dishonouring nonsense *worse* by insisting he will torture and burn people for all time. Forever. Never ending suffering inflicted somehow somewhere on someone other than me because I believe words written in a book. As a teaching purporting to be from God it is

A Lie

A Fiction

A Fabrication

A Hoax

A Myth

A Forgery

A Bribe

If we heard of a father being arrested and charged with the sickening—crime of torturing and roasting his seven little children's flesh for seventeen years—, they being punished for infractions and disobedience to his wishes. . . . what would you feel and think? If you attended court to hear his wife justify this barbarous cruelty and their children testifying through torrents of tears as they removed clothing to reveal shocking scars on each of them from the eldest (who reported him to the police) to his youngest girl who was just three years old; what would you think and feel?

Yes, those children trembled when their father spoke. On occasions privately and uncontrollably released excrement in terror of 'a roasting' on their innocent little bodies when accused by their 'loving father' of being disobedient! The complicity of the mother in this savagery was complicated by her fanatical loyalty to her husband who she sincerely believed had the 'right' to punish disobedient children in this horrendous uncivilised and illegal manner! Some members of this jury (6 men and 6 women) wept silently as each child testified to torture inflicted by its parents. As the last child revealed seven scars she vomited on the stand and the judge called a 15 minute recess to comfort the child and have the court cleaned. It was during this short break that knowing glances and whispers passed from one to the other. As the court reconvened, the defence sat devastated and the forewoman of the jury asked the judge not to send them out to consider the verdict as it was a foregone conclusion. The judge concurred and spoke, "In my 50 years practicing law this is the worst case I ever had the displeasure of listening to. You the parents entrusted with care and protection of seven totally innocent children betrayed them in a most heinous manner possible!" (The parents sat impassive as people always do who believe they have done no wrong!)

"I forthwith remove the children from your monstrous care, make them wards of court and instruct the relevant state institutions to protect and keep these long suffering children in a home with nurses and others to help their growth and healing until each one becomes adult and realises what happened to them never happens to other children who have loving parents" . . . He then addressed the parents: "I am aware your defence pleaded for understanding due to mitigating circumstances to the affect you both unfortunately were reared in a religion that brutalises the hearts of their adherents with this God dishonouring doctrine of a literal eternal hellish fire for non believers! He told me you both attended **16** meetings a month, that's **192** a year and you were married at **30**: That means you had heard this terrifying doctrine preached and spoken **5,760** times. Also you were further burdened by the idea that this ferocious fiction was actually ordained by a loving God as written in your holy book which in your deluded minds equates with reality; there is no excuse acceptable to this court, this jury or this public gallery! Your behaviour was quite monstrous and I sentence both of you to 15 years hard labour."

Voltaire said; "If we believe absurdities we shall commit atrocities." The purpose of this allegory is to expose one of the most outrageous tenet's of quite a number of organised disasters called church's, congregations, faiths, et cetera. Ignorant savage religious leaders in ancient times soon learned to capitalise on mans terror of thunder,- lightening, fire, tragedy and the seasons. The sun was both a god and fire and it could provide or take away heat, life, growth and prosperity. It all depended on obedience to the Sun's representatives with appeasement ceremonies and materialistic offerings to priests.

In time, bombastic rantings morphed into traditional liturgies sanctified by 'sacred writings,' scrupulously written and copied by raving clerics. Naturally the insertion of punishment by fire, being the most terrifying and inducing towards conformity (and wealth for priests) crept into various religious texts and percolated down to the frightened, uneducated minds of parents and then to innocent hearts of bewildered children. A belief is a difficult thing to extirpate from the adult mind when it is hard-wired in from childhood and sanctified by words from a 'sacred text' that millions have come to believe most deeply and sincerely is *'the literal word of God.'*

There are seven incontrovertible reasons why this fiction is false.

Science reveals the dead are nonexistent.

An ancient Jewish scroll confirms that scientific fact by stating "The living know at least that they will die, the dead know nothing, their loves, hates, jealousies, these have all perished."

Flesh burns for hours until gone, not years!

It cannot be located underground.

It is the opposite of love.

It teaches people to hate and fear each other and God.

No loving Father could ever contemplate such savagery for his children.

What else need be said . . . Oh yes some Bible besotted, bilious bearers of "good news" will go ballistic and barrack with apostolic apoplexy and apodictically 'prove' from books that "hell is not hot enough" for individual, invidious infidels like me. And of course the Immans for the Koran shall gleefully join the cacophonous chorus of civilised, chivalrous, sincere defenders of the poisonously indefensible. . . .

Ask the following seven questions:

Would I torture and burn my "enemies" for years?

Could I burn a cat?

How do I feel when I hear of people who do?

Does torture desensitize loving feelings?

Would I create hell if I were God?

Does 70 years of error deserve eternity of terror?

Would I be genuinely happy to discover that in fact there never was nor ever shall be such a vile future for even one human being . . . ?

Therefore if the dead are asleep in cemeteries (the Greek word for sleep), is there going to be a resurrection back to life? If there is a God then, yes, of course, it's possible!! And if there is no God then impossible!! It appears from Moses's book Genesis, we were made to live not die. According to the story, a new administration by a world government is coming which God originally intended shall indubitably and irrefragably commence at some chosen point in the future, regardless of what I write, or you the reader believe . . .

It makes sense of the nonsense all around us. I believe it and I could not care less if no one else does. . . ."Hold fast to dreams for if dreams die, life is a broken winged bird that cannot fly." Langston Hughes—American Poet (1902–1967)

Therefore in harmony with reason—common sense, love and the obvious kind character of our creator—Hell is a demonstrable toxic *FICTION.*

5
MAGIC AND PSYCHICS

The American Heritage College Dictionary Definition of Magic is:

"The art that purports to control or forecast natural events, effects or forces by invoking the supernatural."

"The practice of using charms, spells, or rituals to attempt to produce supernatural effects or control events in nature."

"The charms, spells and rituals so used."

"The exercise of sleight of hand or conjuring for entertainment."

"A mysterious quality of enchantment."

"Possessing distinctive qualities that produce unaccountable or baffling effects."

When a good magician performs a brilliant trick eyes are deceived into believing fantasy is reality. Hearts are charmed and minds suspend logic. However, in time we come to accept that no matter how good magic is, it's still mostly "sleight of hand" with its roots in deception. Each magician knows this, just as surely as Egyptian priests knew the dead were truly dead and the invention of something unseen (like the Emperors new clothes) they called "KA" which was, they believed, the force that animates us and thought people lived on eternally from death ... in a so called spirit realm.

The ignorant priests were so good at teaching this fiction that virtually every wealthy Egyptian spent more money on his tomb than his home ...

When Dubliner Bram Stoker wrote his best selling fantasy *Dracula* (which means "Son of the Devil") in 1897, he may have been aware of the ancient Egyptian, *Book of the Dead* which had drawings of half-man/half-animal depictions which have been found in many tombs. These fictional "undead" creatures left their coffins in the early mornings supposedly, and returned at sunset after mingling (it was thought) in the world they had known and been familiar with.

Bram Stoker reversed this concept and had Dracula leave his coffin at night time to frighten young vulnerable women. In the movie he is cleverly given wings with which to fly and so haunt and terrify his victims before he sinks his fangs into beautiful female necks to suck out their fresh blood as they usually lay innocently sleeping alone in their beds. It is not surprising this fiction is one of the most celebrated tales of terror ever written, ranking in sales alongside best selling books like Charles Darwin's *Origin of Species, The Koran,* the Harry Potter series and many others to numerous to mention.

Priests and witch doctors have always thrived in ignoble cultures of ignorance where arrogant presumptive pronouncements became the panacea for all questions and ailments. However, today because of science, travel, education and reason, we are not so gullible. It is precisely because of these factors that we can assign fictions and fantasies to where they belong—the magician's box of tricks! Once we do that we are free from the peddlers of putrid pseudoscience.

Or possibly he might have been familiar with Greek mythology where the sea god—PROTEUS—could change his shape at will!! (A favourite sleight of hand trick coupled with time and used by Evolutionists to assume one species could change into another!!!)

PSYCHICS

Psychics are individuals who pretend to know the future, prey on ignorance, specialise in guesswork, promote nonsense and tend to adopt a supercilious air of esoteric certainty. A medium is similar in their pretence to communicate with the dead. Once it is scientifically established that dead, decomposed people are no people and the (understandable fiction) idea that, well, really they are still living, somehow, somewhere someplace is demolished, then innocent people who miss their loved ones shall never again be fooled by sophist, specious, spellbinding superstitious syrupy spurious shticks.

Superstition is always, both now and in the past, a badge of ignorance, injected innocently by people deceived by gullible others and so on back to various origins. Psychics know this (if not instinctively) at least on a subconscious level and so mountains of wealth is created worldwide on valleys of despair.

An Example: of prayers to the following:

Coca Cola bottle

Campbell's soup can

Tree

Bird

Fish

Cat

Statue

I have chosen the Cat. (They were worshiped in Egypt)

If one million people pray to that cat, on average 10% will get what they prayed for!!! It's known as the law of averages.

Seven things most prayed for:

The sick to get well.

Girls to get good husbands.

Men looking for good wives.

Good Health.

Healthy babies when pregnant.

Healthy children when first married.

Prosperity.

Out of one million people, one hundred thousand approximately shall get what they prayed for *to the cat!* Did the cat answer their innocent/selfish prayers? No, of course not.

The inexorable law of averages kicked in with pleasing results for the recipients! If there is a God and the balance of probability coupled with "Beyond a reasonable doubt" establishes this premise. Anthony Flew accepted atheism for 50 years and recently came to the cognitive view as written in his recently published book. *There is a NO God,* published by Harper Collins. The following quote is from Francis S. Collins, New York Times Best selling author of *The Language of God.* "In his youth, atheist, Antony Flew committed to the Socratic principle of, 'following the evidence wherever it may lead.' After a lifetime of probing philosophical inquiry, this towering and courageous intellect has now concluded the evidence leads conclusively to God. His colleagues in the church of fundamentalist atheism will be scandalized by his story, but believers will be greatly encouraged, and earnest seekers will find much in Flew's journey to illuminate their own path towards the truth. The next obvious question is does He answer prayers? The logical answer is NO. NO. NO. The mystics, priests and preachers are compelled by lack of evidence to proclaim to those who did not get prayers (bribes!) answered that it is due to 'Lack of Faith!' Bullshit I say!

For example, if I had a child born blind and I asked God to cure him/her (which I assuredly would not!) and for argument sake the child was cured!!! Then my deep respect for my science genius architect creator would immediately diminish! Why? God would then be partial because he cured my child and left many others in miserable darkness! No prayer is not answered at this point in mans' dangerous history.

If a person visits a psychic who pretends to conjure up the dead, it is a dastardly, damnable, deadly lying subterfuge. A bereaved person who is ignorant of the truth will naturally seek succour in hopeless hope, happy in the conviction that something magic has transpired in a dark room. (Paid for by the bereaved)

If all the magicians, psychics, dervishes, witch doctors, priests and preachers on the planet meet together and pray in earnest for a peaceful world, it *is not going to happen!* Peace hopes are spurious in rationale and unworkable in reality. That is the power of truth and reason. Praying for goodies simply sets people up for delusional disastrous desultory disappointments . . .

I am not saying 'do not pray'—I rarely do. What I am conveying is do not set yourself up to be let down. If you believe there is a creator then at times when it feels natural and you are enjoying life, expressions of gratitude might be appropriate.

The fact is nobody can predict the future! Intelligent guesses are anyone's guess and the reality is we all would like to know the future. It is knowledge of this basic human desire that feeds the need of greed that bleeds innocent human hearts, pockets and rockets the charlatan professors of claptrap to the banks.

"Change your thoughts and you change your world." Norman Vincent Peale—American religious minister—Author '*The Power of Positive Thinking.* -' (1898–1993)

Magic and psychics is, was, and, in the foreseeable future, shall remain a deliberate hoax and FICTION.

6
HOLY BOOKS DICTATED BY GOD

Blind faith is an insult to intelligence, an affront to reason, an excuse for thinking and a cop-out when confronted by facts. Faith in fictions is bad enough but when anchored by blindness, he who possess it is a victim twice over. Firstly by social and historical circumstances and secondly through innocence and ignorance. A simple example: We in the western world listen with indifference sprinkled with pity whenever we hear Muslims and their Imams talking about the Koran and how it is some sort of guide for descendents of ancient desert tribes. Our nonchalance is amused by the thought that so many are seriously influenced by so little. For those of us who have read the Koran (and it is difficult to read: I have read J.M Rodwell's translation) we gain some understanding into Mohammad's 7th century Arabian mindset. However, the fact is any one of us born into a village in Saudi Arabia then or now would think and sincerely speak as a convinced Muslim! What does that tell us? It reveals seven notions which we would believe to be true.

The Koran would be believed.

Mohammed would be our guide.

Allah our God.

Life would be meaningless without Islam.

Every other religion is false.

To question Islam is outrageous.

To convert to another religion deserves death.

The same principle can be applied with equal validity to Hindus, Catholics.Protestants. Jews. Mormons, Suffis, Shi-ites, Buddhists, et cetera. . It can even be applied to atheists and agnostics since they have their "holy books" one being the fanciful *Origin of Species* and their modern priests, such as Richard Dawkins. So the question needs to be asked: Why does the conviction and consequent sincerity run so deep in so many on so shallow a premise? The answer is always the same: No child says yes to any religion: Babies are born unbelievers and are influenced by family and culture to believe any opinions foisted on them as normal!

If I read a non-fiction book today, and there are many excellent ones to choose from, I do so from my own volition and decide what is true or false, speculative or factual, imaginative or boring, interesting or ridiculous, savage or kind and so on. However, if I am informed that everything written in its pages is "inspired" and I must believe it to have happiness in my life or any hope of a decent future, the book shall immediately become suspect and its contents taken with a pound of salt! And yet this is the very premise that virtually every so called holy book is reverently postulated! Why? Tradition and a misguided respect for the written word, coupled with pontifical impudence are put forward as reasons why we must accept their particular views!

There are seven reasons why we must accept and believe in their entirety "Holy Books"

Tradition!

Inspiration!

Rewards!

Threats!

God's pronouncements!

Guide for living!

Revelation!

Seven Answers:

Tradition: The slow accumulation of ideas fossilised by time and sanctified by ignorant priests, elders, bishops, imams, ministers, rabbis and others.

Inspiration: Ideas recorded which may or may not have moved individuals at certain times to behave in particular ways! It is this notion which when believed can motivate the kindest of men to behave in the cruellest of ways towards fellow human beings of different childhood persuasions and ALWAYS sanctified by certain relevant "holy" texts! How else could it possibly happen that sane, kind, religious people could behave with such gruesome cruelty unless they had 'inspired texts' to sanctify horrific savagery? Conscience is ignored, human feelings suppressed and love denied.

Rewards: These must be for non-thinking masses otherwise they face the loneliness of a future with no rewards! They rarely ask: "If there was no heaven. Nirvana or paradise would I still accept the notion that my holy book is actually from God or written by mortals!" Only you the reader can answer that.

Threats: This is one sure device that lets you know it's from men who wish to kill or silence opposition. Sadly, one only has to read the origins of mankind as recorded in Genesis or the beginnings of Islam as written in the Koran or the crusades to see the savage heart of man!

God's Pronouncements: Why write about this when everyone who is anyone who knows anything about 'Gods words' knows for a fact they are 'the Saved' and people like me are 'Damned.' To remind people, nobody knows 1% of one millionth of anything only generates anger about everything and so they fall back on Allah's, God's , Yahweh's or the Lords eternal "pronouncements" and are soothed. Ahhh how nice!

Guide for Living: It's embarrassing for religionists how others with different beliefs (or none) from mine, appear to live good blameless lives. Anyway, if people adopt our code in our book the world would be a better place! This is the reasoning of good people with bad ideas only they don't know how bad that one is! They are oblivious to the fact that good people do good things because they are good and not because bad behaviour is outlawed in a book. When a bad person does something good that does not make them good and when a good person does something bad, that does not make them bad. Our guide for living throughout history—today and forever is: our moral conscience . . .

Revelation: this is perhaps the most contentious for in its etymology it is suggestive of "something revealed by someone divine." Because I understand through science, positivism, reductionism, empiricism and probability that there is a (Cognitive Artistic Genetic Engineer) or God, I state:

Possibly mankind might have had a history with less bloodshed and cruelty if written records of mans' miserable existence had never been recorded!

However we do have those records and each person should decide for themselves what might or might not be true bearing in mind we can **NEVER** ever be truly certain of statements and actions of humans written thousands of years ago. Nevertheless of one rock solid fact we can be certain—GOD WROTE NONE OF THEM AND TO SAY SO IS A DELUDED FICTION! OK . . . Yahweh possibly wrote on stone the civilizing ten commandments.

I have read all the major books purporting to reveal God's (or gods'!) purpose for mankind and virtually all are dreary dreadful dross! However, there seriously appears to be one genuine exception. Nevertheless, before I reveal which one, let me state that I do not now nor ever shall at any time in the future accept in the embarrassing manner of fundamentalists that these collected historic documents are inspired with the significance that adjective holds in the minds of various conflicted confused certified rabid religionists. . . .

Seven Reasons why I reject "Inspiration"

It license's interpretative animosity.

It confuses opinions with facts.

It authorises religious cruelty.

It produces pharisaic personalities.

It sanctifies sociopathic savages.

It legitimises murderous mayhem.

It diminishes a loving God.

The document I specifically refer to was written about 3500 years ago and in plain language reveals man was born to live not die. Something went wrong 240 generations ago and today a suffering, groaning mass of confused humanity exists. We are born, live, laugh, grow, become sick, old, cry and die. However, there does appear to be an interesting alternative for a future peaceful world under an administration of the legitimate owner of planet earth. If you the reader wish to know more then read Genesis in the Hebrew book and two parts of the Greek writings called *Luke and Acts.* (Note: I don't care if you do or don't read them!)

Claiming with certainty that a book is from God is an understandable construct in the minds of people reared to believe such a proposition. However regardless of sublime teachings, or how it may have changed us to strive to be better people or give us hope for the future, the fact is we can never be certain and fighting for that uncertainty diminishes us.

Respect the good in it and reject what you honestly in your heart cannot accept and leave its origins and meanings to God!

Until the advent of printing around 500 years ago the average person throughout the 5,500 recorded years of history had no access to books and even if they had, probably could not read. Printing coincided with the greatest schism in Christendom and facilitated fanaticism. People on both sides died for beliefs whose origins they were ignorant of and martyrs 'for the cause' were born from blood. The tragedy in all this was an ignorant legacy on both sides. New generations born were bred to believe the unbelievable unhealthy notion that every word on every page in every generation was the actual words from the creator!

7 Possibilities

God's message—possibly!

God's word—maybe!

God's laws—some yes and others no!

God's ways—yes and no!

God's advice—sometimes!

God's purpose—assuredly

God's personality—yes and no!

The above is not to diminish the soaring logic of purpose, poetry, beauty, wisdom, understanding, history, prose, morality, sense, guidance, hope, happiness, revelation, enlightenment, values, wonder, mystery, et cetera, and realistically—madness and murderous mayhem as recorded in the greatest book on earth, the Bible. It also records an astonishing story about the greatest man who ever walked this planet, Jesus, the amazing Jewish prophet and rabbi.

This extraordinary young man when asked, what appeared to be a trick question in relation to his Jewish religion's 613 laws (codified by Maimonides—1135—1204 A.D) "Teacher what is the greatest commandment?," answered, "Love God and love your fellowman." So why fight over words, statements—texts—letters, meanings, when it all boils down to giving credit to the creator for life and doing to others what we would like them do to us.

In conclusion I recently read an interesting comment which said in essence; If God wanted his children to fight and squabble among themselves the one sure way would be to throw them a book and the idea that he wrote it . . .

Ideas like beliefs are levers for certain types of behaviour. For example a fundamentalist's beliefs and ideas by definition have him/her hardwired and programmed to condemn all others who do not "see" things their way.

They must condemn those who oppose and any punishment handed down is sanctified by relevant "inspired" cruel texts. The tragedy of fundamentalism is it is fundamentally flawed in its sacred roots and its poison is social exclusion to the detriment of all infected.

A book may have beauty – truth – science – poetry – good laws – excellent moral precepts and give meaning to life. It may have all of these and still not necessarily be directly from God. It is for each person to decide honestly for themselves what they believe and what they reject.... That surely is the premise on which all honest belief's rest.

7
EVOLUTION

Evolution can be likened to a train with frightened passengers passing through towns called: Guess, If, Possibly, Asserted, Suggested, Conceivable, Might, Chance, Probable, Theory, Assumption, Viewpoint, Apparently, Hypothesis, Supposed, Ad infinitum.

The innocent unfortunate human beings on board are viewed by their "superiors" as species of a sub-human variety to be cruelly murdered upon arrival at the terminus known as 'Hell on Earth' . . . *AUSCHWITZ*. For years this savage sanctuary to Nazi idiotic ideology became the violent graveyard for hundreds of thousands of innocent men women and children. (Imagine!)

In this morally polluted, poisonous, terrifying atmosphere . . . *Survival of the Fittest* was practiced to a savage degree! A formula much worse was undertaken at each and every miserable train's arrival . . . NATURAL SELECTION . . . Here was the egregious epicentre of epileptic madness! A written lie was posted in German above the entrance gate: It translated: "Work Makes You Free" The originators of Nazism were cruel sophisticated liars, who actually had organisational backing of Christendom's clergy and felt (knew!) they could activate "the final solution" by murdering tens of millions of 'sub-species' when victory was theirs! In the meantime 'ethnic cleansing' which to one degree or another, corrupted the life of all who engaged in it, continued in conquered lands. If the Nazi's had an ounce of courage they would have removed the lie and posted "NOT GERMAN—THEN VERMIN!"

This is what they were groomed to believe for almost twenty years since the publication of a fictitious, non-fiction: *Mein Kampf*, (My Struggle) by a heartless intelligent, cruel, purposeful monster: Herr. Adolph Hitler.

The theory of evolution is an observational practice which cannot and does not reveal origins. Anything an evolutionist states about origins is speculative guesswork and often times pure fantasy! Charles Darwin was a gifted naturalist with a penchant for taxonomy. In his "one long argument" extolling his theory as written in his monumental damaging detonation of sincere convoluted convictions, entitled *Origin of Species* not once does he mention the recently coined word: Evolution: The title suggests in its descriptive word "Origin" that the following 384 pages reveal plants—animals and mans' origins! NOTHING could be further from the truth!

His interesting short introduction of five pages has 10 suppositions. On page 20, Chapter I, titled "Variation Under Domestication," there are no less than 15 suppositions, i.e. "I think, I presume, apparently, probably, my impression, appears to be, seems probable." By the end of page 73 there are, 184 suppositions, by page 146 there are 631 suppositions, and by page 301—1,255 suppositions. Finally on the last page of his science fiction fantasy there are 1,515, yes ONE THOUSAND FIVE HUNDRED and FIFTEEN SUPPOSITIONS! I mention these statistics not to weaken his argument (which it does in consequence) but to strengthen his taxonomical treatise of a very weak theory masquerading as fact. The result of this despairing conclusion is: The universe is a cosmic accident produced by millions of years of random explosions and mutations: for no reason, no purpose, no point, no joy, no future, a cosmos where human beings have no more meaning ultimately than a white round pebble found on a beach! (Pagan Romans gave significant meaning to this: when they gave a white pebble to someone its meaning was friendship forever, solid as a rock!)

In *River out of Eden, which presents a Darwinian pointless view of life,* written by Richard Dawkins (1995) whose intellect I admire and whose conclusions I am amazed at characterised the universe as having "No design, (WHAT?) No purpose, no evil and no good, nothing but blind pitiless indifference." He and his acolytes have erected a substitute quasi-religious edifice with similar insidious, insulting, intimidating, intellectual consequences!

Those who question the trinity of:

Charles Darwin   *(The Father)*

Evolution   *(The Son)*

Scientific speculation   *(The Holy Spirit)*

are pigeon-holed as mental monkeys, unscientific, not bright (Brights!) heretics, backward, etc! However the opposite is actually the case. I say as Louis Pasteur wrote: "A bit of science distances one from God, but much science nears one to him" Dr. Wernher von Braun: (1912–1977), NASA director and 'Father of the American Space Programme' wrote,. "Manned space flight is an amazing achievement, but it has opened for mankind thus far only a tiny door for viewing the awesome reaches of space. An outlook through this peephole at the vast mysteries of the universe should only confirm our belief in the certainty of its Creator." . . . "It is in scientific honesty that I endorse the presentation of alternative theories for the origin of the universe, life and man in the science classroom. It would be an error to overlook the possibility that the universe was planned rather than happening by chance." . . . "Atheists all over the world have . . . called upon science as their crown witness against the existence of God. But as they try, *with arrogant abuse of scientific reasoning,* to render proof there is no God, the simple and enlightening truth is that their arguments boomerang. For one of the most fundamental laws of natural science is that nothing in the physical world ever happens without a cause. There simply cannot be a creation without some kind of spiritual Creator . . . In the world around us we can behold the obvious manifestations of the divine plan of the Creator."

The question needs to be asked: How and why did this gifted taxonomic naturalist who specialised in observational conclusions have such an impact, in particular, on the western world? A distinguished university professor of Biology at the University of Massachusetts, Ms Lynn Margulis, wrote as recorded in *Dar-*

win's Black Box, Page 26: "History will ultimately judge neo-Darwinism as a minor twentieth-century religious sect within the sprawling religious persuasion of Anglo-Saxon biology." At one of her many public talks she asks the molecular biologists in the audience to *"name a single, unambiguous example of the formation of a new species by the accumulation of mutations."* Her challenge goes unmet. Proponents of the standard theory, she says, "wallow in their zoological, capitalistic, competitive, cost-benefit interpretation of Darwin—having mistaken him . . . Neo-Darwinism, which insists on the slow accrual of mutations, is in a complete funk."

There is a conspiracy of cowardly silence among many scientists who know in their minds and hearts that something is wrong. In 1859 when *On Origin of Species* was first published England was approaching its zenith of empire.

Europe spent 1,554 years cursed with religious dissensions since the Roman emperor Diocletian's time (284—305 A.D.) who strengthened his authority by becoming a semi-divine ruler and making his palace 'a holy house.' The attitude of people towards Christendom was one of unquestioning, abject fawning and obsequious subservience. Religious certainty, hypocrisy, pretension, convention, tradition and societal cruelty were the historic offspring from a corrupt edifice.

This explosive concoction sinking under the groaning mass of human misery and destitution simply needed a match to ignite the pent up rage in millions of the world's starving ignorant serfs. Darwin's book appeared to give intellectual dynamite to demolish clerical control of almost every aspect of life! In his book, *Deep History and the Brain,* Daniel Lord Smail credits Stephen Jay Gould for calling, "The discovery of deep time a cosmological revolution of Galilean proportions. The reality is chronological certainties came crashing down in the 1860's with the sudden and wide-spread acceptance of geological time."

Science was unleashed from the ludicrous laughable certainties of Christendom's crucifying orthodox traditions. The age of reason was alive and flowering in all its empiric beauty from the dead dung heap of theological treacherous tyrannies, which had hounded, hunted, burned and brutalised any who dared question their invented fictions, for almost 2000 years.

This "valley of tears" for tens of millions of suffering humanity living in abject poverty and destitute of hope barely registered on Europe's ruling elite and the growing educated middle classes. The world's love affair with clerical officers was about to collide with the explosion of science, knowledge and reality.

Russia was mired in shocking poverty for most of her people and life was brutal, often violent and a ticking time bomb waiting to go off. Simon Sebag Montefiore recently wrote *Young Stalin*. This fascinating book powerfully demonstrates how violent men desensitised by shocking social conditions, corrupted by ignorant priests, demoralised by amoral "Royal Families," brutalised by cruel servants of a cruel state could through sickening violence take control of a terrorised nation. The ultimate tragedy is Stalin became an atheist (he used to be a student priest) and savagely put down any opposition. He was directly responsible for the sad deaths of millions of innocent Russian people.

From 1860 to 1960 there was a staggering increase in communications, methods of travel, inventions, education, science, military madness and the creation of weapons of such destructive magnitude that many students of history currently see an inexorable inexplicable insensitive inhuman drift towards doom!

When tens of thousands of scientists, philosophers, educators, among others in the late 19th century were aglow with progress and pondering mans' place in the universe along comes a book in 1859 called *On Origin of Species* which speciously appeared to give some theoretical support to the novel idea of molecular "modification through natural selection" meaning, as Darwin imagined, that one species slowly changed into a different species. Reader note this well: *it does nothing of the sort*. The slow (or fast) accumulation of toxic mutations to produce a new species, never happens—is not happening now nor shall ever happen in the future. This is a scientific FACT . . . (as Darwin in his molecular ignorance imagined.)

However, I am open to persuasion through:

Evidence.

Facts.

Logic.

Science.

Reason.

Teleology.

Empiricism.

I esteem Charles Darwin's observational eye in ascertaining habits and instincts of fish, birds, insects, animals, plants, flowers, among many other living things and I respect his vast taxonomical categories in each discipline . . .

Educated intelligent men and women began, in droves, to question everything. Philosophers and thinkers like Socrates, Pascal, Voltaire, David Hume, Thomas Paine, plus hundreds more began to be read and quoted in various publications. John Stuart Mill, the English philosopher and classical liberal thinker (1806–1873) wrote about his father, in *The Autobiography*; "His aversion to religion, in the sense usually attached to the term, was of the same kind with that of Lucretius: he regarded it with the feelings due not to a mere mental delusion, but to a great moral evil. He looked upon it as the greatest enemy of morality: first, by setting up factitious excellencies—belief in creeds, devotional feelings, and ceremonies, not connected with the good of human kind—causing these to be accepted as substitutes for genuine virtue: but above all, by radically vitiating the standard of morals; making it consist in doing the will of a being, on whom it lavishes indeed all the phrases of adulation, but whom in sober truth it depicts as eminently hateful." Taken from Christopher Hitchin's interesting book—*God is not Great* page 15 under the title, 'Religion Kills,' It was no wonder the Pope called Vatican I in 1869–1870 to try and stem the treason of reason percolating throughout his sad diminishing realm. But it was too late, the genie was out of the bottle and the glass was smashed. Never again would obscurant obtuse religious dictators of diabolical doctrine befoul common sense with ludicrous mysterious abstractions. Never again would educated people be afraid to question statements from "holy books." Never again would people believe in a sickening raging burning hellfire for dissenters or unbelievers, a doctrine originated to terrify and convince people of Gods cruelty with the consequent result of poisoning believers with a deep seated and understandable hatred for such a despicable dastardly deity. Imagining something as true is vastly different from knowing the truth *based* on evidence reason and reality.

For 150 years Charles Darwin's book has provided the lever (belief) that animals, vegetation and mankind originated from zero."Ex nihilo nihil fit." (from nothing comes nothing) Paradoxically he believed in God and is quoted as saying: "I believe life was originally breathed in one or many species from which we are descended."

However, this dyspeptic dystopian like many aristocratic English country 'gentlemen' who visibly exhibited elaborate manners to their peers yet seldom extended them to the groaning working class and rarely while shooting on their luxurious estates. In the excellent biography by Cyril Aydon, *Charles Darwin—His Life and Times,* page 19 is revolutionary in revealing his attitude to killing birds: It states in paragraph 1, "Of the fever which shooting aroused in him in his teens," he was later to write,

"I do not believe that anyone could have shown more zeal for the most holy cause than I did for shooting birds. How well I remember killing my first snipe and my excitement was so great that I had much difficulty in reloading my gun from the trembling of my hands."

I am personally appalled at such indifference to creature suffering. Most bullets miss or wound, so how many birds died slow agonising deaths to give "pleasure" to the landed gentry? His quote is indicative either of his temperament or of his generation, who knows!!! However the social, psychological, political, and scientific impact of his book has undoubtedly been incalculable. In 2007 the Science Channel named its top 100 scientific discoveries of all time, and trumpeted as number one, Darwin's theory of evolution! America was discovered in 1492 by Christopher Columbus. Louis Pasteur (1822—1895) discovered how to vaccinate against rabies and during his lifetime saved almost 20,000 victims from certain death. Sir Isaac Newton (1642—1727) discovered gravity and his three great laws of motion, which underpin all modern physics. I could list 69 other great scientific discoveries which are tangible, observable, and whose provability is accepted as demonstrable fact. However, it must be asked: "What exactly was discovered?" An un-testable, un-provable idea is not a scientific discovery by any stretch of a pseudoscientific imagination! In fact in reality in truth and in print his book has more semblance than substance. (It has in total 1515 SUPPOSITIONS—I know because I counted every single dreary one.)

Yes he "discovered" breeding plants, pigeons, dogs and a lot more experiments besides, showing different strains within the same family become observable over time! Nevertheless any botanist, pigeon or dog breeder could have proved that to him in 15 minutes! However no rose ever became a lily! No pigeon ever became an eagle! No dog ever became a horse! And the fossil record proves that as there are over 250,000 species of plants and animals on view today in museums around the world, not one, I repeat, not one is demonstrably in an intermediate stage!

Did Charles Darwin put flesh on the bones of a brilliant discovery that has benefited mankind for 150 years? No! Did he discover the answer to even one of the seven great questions of all time? No!

QUESTIONS ANSWERED BY EVOLUTIONISTS

How did the universe originate?   By chance!

Is there a God?   Apparently not!

Why am I here?   No good reason!

What is my purpose?   None!

Why must I die?   Because we do!

Where do I go when I die?   The graveyard!

Why is there injustice and pain?   *Because life is a struggle!*

The forgoing seven answers may be simplistic but they encapsulate the bottomless pit of evolutionary, cruel, insignificant meaninglessness that anchors the mind to potential despair.

As a knowledgeable naturalist and sentient sensitive thinking man, he was tormented by life's tragedies, inner contradictions, random cruelties, and mans' sudden dramatic violence, wars and destructive behaviour. He instinctively knew religion could not provide truly satisfying intellectual answers to our brief lifespan on earth.

Although Darwin's idea mostly grew by slow accumulation, there was one "Eureka!" moment, when he read 'An Essay on the Principle of Population' by Thomas Malthus. In this book, written in 1798, Malthus argued that populations- both human and animal- will always multiply until they exceed the amount of food available, at which point the population will crash, only for the purpose to start again. Darwin was excited: "It at once struck me that under these circumstances favourable variations would tend to be preserved, and unfavourable ones to be destroyed. The result would be the formation of a new species." (Now proven to be false.). From—*The Great Scientists* by John Farndon and published by Arcturls Ltd in 2005, pages 91–92.

It is hardly creditable that this searching man could have after all those years of diligent study, settled for so little, in representing in his mind, so much! But he evidently did! Famines or other reasons for lack of food and the tragic consequences have been with us for centuries. But the opposite of what Malthus claimed and Darwin accepted is actually the case. Populations have inexorably increased for millennia and particularly since the 19th century, so the lever for his theory is in fact false! Yes, of course, entire families and communities perished in times of food shortages. Some 168 years ago in my country, Ireland, 3,000 people a day perished from a horrible famine. However, Ireland is now one of the most vibrant and prosperous countries on earth (As I write in the summer of 2008, but hey, who knows the future!)

So once again why has a book published 150 years ago continued to exert an influence out of all proportion to its intellectual and evolutionary weight? I use the adjective "evolutionary" here in its correct context, for example the evolution of soccer, boxing, music, medicine and thousands of other subjects. Here it fits because our minds can conceive of the origin of soccer, plus we have written evidence of its history and photographs to substantiate realistically its inception and growth in England, for example.

However, if one hears or reads, "When we crawled out of the sea millions of years ago," our minds do a triple flip simply because we cannot imagine when man/we were supposed to be "creepy crawlies" eating fish and slugs!!! Or we may hear "when we climbed down from the trees millions of years ago!" our cognitive reasoning powers flip again because we cannot visualise actually living up a tree! (Try climbing an oak and living there for seven days. Try!!! It's ludicrous!!!)

There are seven reasons why a proven fiction-evolution has evolved as a quasi- religious belief system which is now hawked about by talented writers and zoologists like Richard Dawkins, Christopher Hitchens, Sam Harris and Daniel C. Dennett among many other intelligent and capable individuals who have become on the world stage—*"the high priests of the highly improbable foisting the impossible on the impressionable."*

7 Reasons Why:

They do not wish to believe in a creator for various understandable reasons.

They discovered that organised religion is a gigantic hoax.

As children their minds were infected by it.

In university evolution was taught with smug certitude.

Educated people wish to distance themselves from biblical embarrassing fundamentalists.

Peer pressure.

They think there is scientific proof when absolutely none exists.

SEVEN ANTI- EVOLUTION SCIENTIFIC QUOTES:

Britain's New Scientist observed that, "an increasing number of scientists, most particularly a growing number of evolutionists . . . argue that Darwinian evolutionary theory is no genuine scientific theory at all . . . Many of the critics have the highest intellectual credentials."

When a special centennial edition of Darwin's Origin of Species *was to be published W.R. Thompson, then director of the Commonwealth Institute of Biological Control, was invited to write an introduction. He said, "As we know, there is great divergence of opinion among biologists, not only about the causes of evolution but even about the actual process. This divergence exists because the evidence is unsatisfactory and does not permit any certain conclusion. It is therefore right and proper to draw the attention to the non-scientific public to the disagreements about evolution."*

Summarizing some of the unsolved problems confronting evolution, Francis Hitching observed: "In three crucial areas where (the modern evolution theory) can be tested, it has failed: The fossil record reveals a pattern of evolutionary leaps rather than gradual change. Genes are a powerful stabilizing mechanism whose main function is to prevent new forms evolving. Random step-by-step mutations at the molecular level cannot explain the organized and growing complexity of life"

"Evolution is a fairy tale for grow–ups. This theory has helped nothing in the progress of science. It is useless."- Professor Louis Bounoure, former President of the Biological Society of Strasbourg.

"Scientists who go about teaching that evolution is a fact of life are great con-men, and the story they are telling may be the greatest hoax ever. In explaining evolution, we do not have one iota of fact."- Dr. T.N. Tahmisian of the United States Atomic Energy Commission."

"The theories of evolution, with which our studious youth have been deceived, constitute actually a dogma that the entire world continues to teach: but each, in his specialty, the zoologist or the botanist, ascertains that none of the explanations furnished is adequate. It results from this summary, that the theory of evolution is impossible."—Paul Lemoine, director of the Natural History Museum Paris recognizing the lack of evidence as early as 1937.

"I myself am convinced that the theory of evolution, especially the extent to which it's been applied, will be one of *the great jokes in the history books of the future"*—Malcolm Muggeridge, world famous journalist and philosopher.

A piece of clay, sand, sawdust, ashes or dust could all reasonably be said to have apparently no design but an ocean liner, supersonic jet or submarines most assuredly have, as most normal people would agree. So I say to Richard Dawkins and his overawed disciples from an under-whelmed man "Wake up and smell the coffee" (I know, I know some will want to question is coffee really coffee, is a smell really a smell, is sleep really unconsciousness) However a brief study of the universe—our solar system (system, according to The Universal Dictionary means: "A group of associated stars, planets, or other bodies") planetary orbits, phases of the moon every 29–5 days to name but a few all SCREAM the opposite of Richard Dawkins puzzling conclusion of "No Design" To me it is, Yes, Yes, Yes, planned mesmerising thought provoking architecture . . . Law implies order and order implies intelligence.

If a man (or woman) born into a loving family lost their sight at 10 years of age, then in a tragic accident lost both parents, and two years later lost brothers and sisters in a fire, then had to live in an orphanage unloved for six years before moving out to what would appear an indifferent world! Could we blame him/her for concluding there could not possibly be a loving God! Especially if the religion they were born into taught . . ."God took them to heaven." How sick! How sad! How patently untrue! Supposing that person who once had vision, therefore, understood colours, distance, shapes, speed and seasons, was in a city fifty years later and received a blow to the head by a mugger, and suddenly from that act of savage violence sight was restored! Imagine the joy, astonishment, thrill at being able to see after fifty years of blackness! Imagine that person had a natural interest in astronomy but could not pursue it as a career because of visual incapacity! Imagine by coincidence the blow took place outside a planetarium to the delight of the victim! Imagine the ex-blind person going in and slowly observing the precise mathematical movements and obvious celestial majesty, beauty, order and planned architecture! (Or Design!) Would you criticize that person for thinking "I have a choice; most of my life I have been unlucky through no fault of my own. I have rejected religion as a disease on the suffering human race! I have embraced agnosticism and possibly sometimes atheism in my loneliness, never finding love and loosing the ones I loved as a child! But today my eyes have been opened literally and I am astonished at the celestial solar system. I must choose; God or no God! Hope or no hope!

The scientific mathematical brilliance in front of my eyes tells me something or someone is responsible for this daily night time of astonishing orderly cognitive complexity that people with sight take for granted and loose sight of. If science is "the experimental assessment of reality" then today my fiction is swapped for my reality.

I have been denied the pleasure for fifty years of observing the heavens and I am inside this planetarium for only seven hours, and already my intelligence is on fire with respect for science, disgust for religion and awe for the Cognitive Artistic Creative Architect. Oh, this is the happiest day of my life. Today I have suffered violence, received sight, observed indescribable beauty, realised indubitably there is a God. I am a happy, happy man . . ."

The world Charles Darwin lived in was one of scientific discovery; trains, metal ships, bridges, faster communications, medical progress, intellectual daring, philosophical posturing and religious questioning. However the clergy of Christendom were partners with a pompous and supercilious growing world power, Great Britain. Into this mix was thrown a 384 page book, the result of many years' observations of plants, insects, birds and animals. In 1859 the literary litany of Darwin's wide rooted shallow roots took hold! For educated, thinking, empathetic people this was the key to unlock the lock that chained them to treacherous theological mysterious intellectual mayhem! A phoney war was declared between:

Science V Religion.

Evidence V Faith (in an un-provable theory.)

Reality V Belief (in an un-testable idea).

Facts V Blind acceptance of a fiction.

Discovery V Speculative postulations.

Creative facts V Evolution.

Scientific demonstrable facts V Darwinian evolution.

The sad reality is many scientist's threw out the baby with the bath water and embraced Darwin's defective analysis of nature and his un-testable and unproven theory as fact. Species crossed with different species are sterile. Species crossed with same species are fecund. However much theorising or speculating is done, those two results are facts! Humans produce humans, elephants produce elephants, cats produce cats, pigeons produce pigeons, alligators produce alligators, banana trees bananas, whales produce whales and the entire cross breeding in the world has not and cannot produce one single new species. There is, through DNA, genes and molecular structures at conception a myriad of variations possible: but no woman has ever given birth to a dog and never will!

Evolution has failed on:

Fossils.

Genes.

Mutations.

Science versus Religion has been cleverly addressed as "two non-overlapping magisteria" but that is attributing too much dignity to both. Science simply means "to know" or as the Universal Dictionary states: "Learning or study concerned with demonstrable truths or observable phenomena, and characterised by the systematic application of the scientific method"

Religion means "to worship" or as the Universal Dictionary states;

"The expression of man's belief in and reverence for a superhuman power or powers regarded as creating or governing the universe. Any personal or institutionalised system of beliefs or practices embodying this belief or reverence."

When I state I practice no religion because my beliefs are scientific, I honour both The Creator and science. For example a myth has been perpetuated since Darwin's day, when Thomas Huxley affectionately known as "Darwin's Bulldog" turned the tables cleverly on the Right Reverend (note the self aggrandising, stupid title) Samuel Wilberforce, D.D. Bishop of Oxford. At this now famous meeting it is reliably reported the following took place as recorded in Cyril Aydon's biography of Darwin, pages 220—223:

"It was into this supercharged atmosphere that Samuel Wilberforce, now launched himself. He was not intimidated by the occasion. He was the son of the great anti-slavery campaigner William Wilberforce, who had coached him in public speaking as a child, and he was a considerable orator. He was also a man of great intelligence, and he was beyond argument the most influential churchman of the age. Ironically, his father had been a friend, and fellow campaigner, of both of Darwin's grandfathers, but the son's intention on this Saturday afternoon was to trample the name of Darwin in the dust.

As well as his intellect, Wilberforce was noted for his ingratiating manner. He now proceeded to use his charm and rhetorical skills to persuade his audience that Darwin's theory was as unscientific as it was irreligious. Judged by style and form alone, it was a brilliant performance; but the longer he spoke, the more certain Huxley and Hooker became that the only ideas in his head were those that Owen had placed there. (Richard Owen was the uncrowned king of British zoology and superintendent of the natural history collections of fossils, at the British Museum.) There was no doubt that he had a large section of the audience on his side. But just before he started his peroration, he made the mistake that was to cost him his moral advantage. Turning to Huxley, he enquired whether it was on his grandfather's or his grandmother's side that he claimed descent from an ape. He rounded off his speech with a resounding declaration that Darwin's theory was contrary to divine revelation, and sat down to rapturous applause. As the noise abated, voices were raised, demanding that Huxley be called to speak. Henslow acceded, and Huxley began his dissection of Wilberforce's performance. He drew the audience's attention to the contrast between the speech's confident manner and its absence of solid matter. His most cutting comment was made in response to Wilberforce's tasteless question. The excitement that followed his speech left those present with conflicting memories of the precise words he used and the exact point at which he used them. But the essence of his reply was remembered by all who heard it. If he were asked to choose, he said, between an ape for a grandfather, and a man of great gifts, who used those gifts for the purpose of introducing ridicule into a serious discussion, he would unhesitatingly affirm his preference to the ape."

The arrogant insulting language of the so called bishop Mr. Wilberforce instantly squandered any moral authority he may have had! Since then some scientists have disdainfully, disrespectfully, disingenuously, discountenanced the courageous outnumbered scientists who have pointed out the scandal yes scandal among the scientific fraternity for substituting theory for fact;

At this point I could give thousands of erudite statements from eminently qualified scientists (and I don't care what, if any, their religious convictions are because they are an irrelevancy: what matters are facts—not fictions) so consider carefully why it was said and does it ring true! However, before I do that I list

the speculative specious spurious languid language of theorists.
(These words are from *Origin of Species*)

Imagine.

Infer.

I think.

It seems to me.

May be said.

Might.

Likely.

Chance.

Conceivable.

May be true.

Assume.

Theory.

I believe.

Inclined to suspect.

My view.

My impression.

Apparently.

Under this point of view.

Appears to be.

Change may I think.

Perhaps.

I presume.

The view suggested.

Seems probable.

Allude.

Almost certainly.

If.

Probably.

Seems to be.

The Doctrine of chance.

May be.

Perhaps.

May have.

Attributed.

Likely produces.

Doubt.

Not knowing.

It could be.

Presumptive.

This view.

Or even.

Variation may be said.

Highly probable.

Opinion.

I am doubtfully inclined to believe.

Might have been.

We may suppose.

I have deliberately not listed the pages these suppositions are on as I would like people themselves to read Darwin's dreary theory, and if anyone manages to reach page 146 you will have read 638; yes six hundred and thirty eight suppositions! Christopher Hitchin's, my linguistic hero, said: "What can be asserted without evidence can also be dismissed without evidence."

I now let the following scientists speak for themselves:

Biologist Michael J. Behe wrote in his book; *Darwin's Black Box (Published by Free press 2006)* © *1996*

"Meanwhile, the cellular black box was steadily explored. The investigation of the cell pushed the microscope to its limits, which are set by the wavelength of light. For physical reasons a microscope cannot resolve two points that are closer together than approximately one-half of the wavelength of the light that is illuminating them. Since the wavelength of visible light is roughly one-tenth the diameter of a bacterial cell, many small, critical details of cell structure simply cannot be seen with a light microscope. The black box of the cell could not be opened without further technological improvements.

In the late 19th century, as physics progressed rapidly, J.J.Thompson a British physicist (1856–1940) discovered the electron; the invention of the electron microscope followed several decades later. Because the wavelength of the electron is shorter than the wavelength of visible light, much smaller objects can be resolved if they are "illuminated" with electrons. Electron microscopy has a number of practical difficulties, not least of which is the tendency of the electron beam to fry the sample. But ways were found to get around the problems, and after World War II electron microscopy came into its own. New sub-cellular structures were discovered: Holes were seen in the nucleus, and double membranes detected around mitochondria (a cell's power plants). The same cell that looked so simple under a light microscope now looked much different. The same wonder that the early light microscopists felt when they saw the detailed structure of insects was again felt by 20th century scientists when they saw the complexities of the cell."

This level of discovery began to allow biologists to approach the greatest black box of all. The question of *how life works* was not one that Darwin or his contemporaries could answer. They knew that eyes were for seeing-but how, exactly, do they see? How does blood clot? How does the body fight disease? The complex structures revealed by the electron microscope were themselves made of smaller components.

What were those components? What did they look like? How did they work? The answers to these questions take us out of the realm of biology and into chemistry. They also take us back into the 19th century.

Commenting on his book in the National Review, George Gilder said:"It overthrows Darwin at the end of the 20th century in the same way that quantum theory overthrew Newton at the beginning"

Palaeontologist Stephen Jay Gould said, "The extreme rarity of transitional forms in the fossil record persists as *the trade secret of palaeontology.*" This devastating honest comment is from *The Panda's Thumb*, 1980, page 181; in other words there is not one missing link, there are billions from the millions of species on this planet today and unambiguous intermediate transitional forms are never—never discovered. So I congratulate him on his honesty.

Writer Christopher Booker, (an evolutionist!) London Times; quoted in the Star, Johannesburg "The Evolution of a Theory" April 20th, 1982, page 19: "It was a beautifully simple and attractive theory. The only trouble was that, as Darwin himself, was, at least partly aware, it was full of colossal holes." Regarding Darwin's *Origins of Species*, he observed: "We have here the supreme irony that a book which has become famous for explaining the origin of species, *in fact does nothing of the kind.*"

Scientist, Stephen Hawking referred in *A Brief History of Time*, to a hoped-for time when an eloquent and unified theory of everything is developed. He says, "Then we shall all, philosophers, scientists, and just ordinary people, be able to take part in the discussion of the question of why it is that we and the universe exist. If we find the answer to that, it would be the ultimate triumph of human reason- for then we would know the mind of God." For some weird reason it appears that humans in general and scientists in particular exhibit an aversion to simple scientific gems of empiric logical wisdom!

Zoologist, Professor William Thorpe of Cambridge University, told fellow scientists, "All the facile speculations and discussions published during the last ten to fifteen years explaining the mode of origin of life have been shown to be far too simple minded and to bear very little weight. The problem in fact seems as far from solution as it ever was."

Astronomers, Fred Hoyle and Chandra Wickramasinghe said, "The problem for biology is to reach a simple beginning and fossil residues of ancient life forms discovered in the rocks do not reveal a simple beginning. . . . so the evolutionary theory lacks a proper foundation!"—*Evolution from Space,* 1981 page 8.

In fact it would take a minimum of 42,700, six hundred page books to try explain the instructions within the DNA of one cell. It is more complex than all the satellite, terrestrial, mobile, land line telephone communications and computer internet connections combined. Yes, just one. And each cell has two hundred trillion tiny groups of atoms called molecules. Newsweek magazine uses an illustration to give an idea of the cells activities, "Each of those 100 trillion cells functions like a walled city, power plants generate the cells energy and factories produce proteins, vital units of chemical commerce. Complex transportation systems guide specific chemicals from point to point within the cell and beyond. Sentries at the barricades control the export and import markets, and monitor the outside world for signs of danger. Disciplined biological armies stand ready to grapple with invaders. A centralized genetic government maintains order." (Newsweek, "The Secrets of the Human Cell," by Peter Gwynne, Sharon Begley and Mary Hager, August 20, 1979, page 48) All this genetic information is systematically in methodically coded sequence and biologically timed arrangement. And it fits on the top of a pinhead!

When the modern theory of evolution was first proposed, scientists had no idea of the astonishing complexity of a living cell. We do now and it is disgraceful, that modern scientists cling to a demonstrably false theory and willingly choose a non intelligent fiction. They can still ignore God if they wish! They can curse God if they want! but the one thing they should not do as scientists is pretend they can prove origins when they know in their hearts through scientific evaluation, that the beautiful astonishing architecture in the universe, solar system, earth species and cells all logically imply a mesmerising architect!

I shall give one more illustration of the mind boggling complexity of something smaller than a pinhead. Before I do, remember you and I once were that size! What does that mean? It means our fathers were producing one thousand sperm per second as young men. Sperm production is the most astounding of the human body's phenomenal production systems. Every day a healthy young male produces nearly one thousand (1000) sperm a second. Each sperm, moreover, contains an entirely unique selection of the father-to-be's genetic material. The sperm are produced in the long, convoluted passageways of the testes known as the seminiferous tubules.

ONTOGENY is the development of a human baby from the second of conception to the moment of birth. (far more interesting than ontology!) In Lennart Nilsson's magnificent book, *A Child is Born (published by Bantam Doubleday Dell Publishing Group, Inc.)* astonishing photographs illustrate the utter impossibility of all this genetic genius, so brilliantly interconnected, each working simultaneously for the benefit of the growing blastocyst, embryo, foetus and baby. "As though at the starting gate of a marathon, 500 million sperm set off toward the elusive finishing line: an ovum concealed in the fallopian tube. Of this teeming crowd, only one can enter the ovum. For odds of 500 million to 1, a life giving prize." page 42, *A Child is Born*

Ejaculation is like the start of a marathon. Each and every single sperm carries in its head the entire male genetic code. So out of 500 million only one gets to penetrate the egg and unite both sets of twenty three chromosomes to make, a new unique human being! To illustrate the complexity and majesty of this "miracle" find a pin and examine its point . . . Now, try grasping the molecular significance in something smaller than a pinhead! Which once was I and you.!

"The human genome consists of all the DNA of our species, the hereditary code of life. This newly revealed text is over 3 thousand million letters long, and written in a strange cryptographic four-letter code. Such is the complexity of the information carried within each cell of the human body that a live reading of that code at a rate of one letter per second would take thirty one years, even if reading continued day and night.

Printing these letters out in regular font size on normal bond paper and binding them all together would result in a tower the height of the Washington Monument." From; *The Language of God* by Dr. Francis Collins. (head of the human genome project and one of the worlds leading scientists—he works at the cutting edge of the study of DNA, the code of life. Professor Collins is a deep believer in the God of Jesus and proclaims it publicly.)

Every second young healthy males produce 1,000 sperm. That is 100 million per day and each and every one has within itself the father-to-be's genetic code. To illustrate the approximate amount of information needed at the atomic molecular cellular level to commence fertilisation and the next forty weeks of miraculous millions of parts of connected flesh, blood and bone all developing according to physiological, mathematical and biological fixed lavish laws.

Observe the following carefully: on the cover of this book is a man's hand. The hand holds a needle and the tip of that needle could contain the entire three thousand million chemical coded letters that turn from a language into a beautiful baby.

This means that precise amounts of crucial genetic information are unleashed every second of every day in every pregnant womb in every country of the world.

What is the point of trying to convince people who have chosen to adhere to a fiction? If there is no God, PROVE IT! If there is a God, PROVE IT! Both contain difficulties of that there is no doubt.

However the evidence for creation on a biochemical level comes down emphatically on the side of empirical facts. Phil

Town's book, *Rule Number One*, page 65 quotes Bosewell in "The Life of Samuel Johnson" (1709–1784) "Mankind have a great aversion to intellectual labour; but even supposing knowledge to be easily attainable, more people would be content to be ignorant than would take even a little trouble to acquire it."

On the other hand, there are well educated people who simply decide to believe the unbelievable: note well the following honest admission from Dr. George Wald, professor emeritus of biology at Harvard and Nobel Peace Prize winner in biology in 1971: "There are only two possible explanations as to how life arose: Spontaneous generation arising to evolution or a supernatural creative act of God . . . There is no other possibility. Spontaneous generation was scientifically disproved 120 years ago by Louis Pasteur and others, but that just leaves us with only one other possibility . . . that life came as a supernatural act of creation by God, but I can't accept that philosophy because I do not want to believe in God. Therefore I choose to believe in that which I know is scientifically impossible, spontaneous generation leading to evolution."

Here we have a retired distinguished professor of biology and a Nobel Peace Prize winner admitting he "Chooses to believe in that which he knows to be scientifically impossible!!!"

Question; would any sensible person want him teaching a science class to our children?

Would you employ an architect to design something for you he knew was *scientifically impossible?*

Would you want a surgeon to recommend an operation on you that he honestly believed was *scientifically impossible?*

Would you set foot in an aeroplane if the captain revealed the engine was designed by an engineer who believed flying was *scientifically impossible?*

Would you continue to go to a family doctor if he revealed that modern medicine and antibiotics were damaging to health and *scientifically impossible* to be of any benefit?

The question is, why do millions of people, educated, intelligent, thinking, and otherwise, cling to fantasies and fatuous fictions? There are 7 explanations:

Childhood teachings

Social influences

Peer pressure

Fear of change

Embarrassment

What others think

Cowardice

If there is a God, there is a heaven—resurrection—paradise, at some future time! If there is no God then there is nothing—no future, no hope, no purpose. I could not care less what a person's inherited religion or chosen one is. Jesus is quoted as saying as he hung up in agony on a Roman implement of excruciating torture. "Father forgive them for they know not what they do!" Whether they knew or not I don't know nor care. The point is they maintained their religious traditions and pretended loyalty to political masters. The casualty of this cruelty was an innocent young Jewish Rabbi. It once again proved Lucretius's observation "To such heights of evil are men driven by religion." (Roman poet—99–55 B.C.E.)

I detest organised religion more than Richard Dawkins, Christopher Hitchens, Sam Harris, Daniel C. Dennet and others. The reason I do is because I believe in a COGNITIVE ARTISTIC GENETIC ENGINEER (GOD) and they don't! The denigration and destruction of God's loving personality through daily disinformation is appalling. I recently read Rick Warrens book, *The Purpose Driven Life (Published 2002—a* magnetic title!) And the cover quote, from Publishers Weekly is: "The bestselling non-fiction hardback book in history." He appears to be a good man in a bad environment. Thirty million copies of this childish book have been sold, mostly to people of a Protestant background. However the nice thing about this is profits from his religious business go to help hungry people in many lands! He has locked in eloquently to the power of love, married to purpose. However on page 300 he laughably and embarrassingly asks us to pray, "to a map or a globe."

I once took my six children to an American preacher who lectured in Dublin on "How God answers prayers and performs miracles" knowing both premises are currently utterly fictitious. At evening's end I asked my children did they see a dead person come to life? Did they note anyone with no legs get new legs? Did they witness a cleft palate child get new lips? No.–No. No. Lesson learned.

The tragedy of good people like Rick Warren is they are forced to accept, believe and shamefully teach a bad doctrine, the most God dishonouring idea ever invented by cold hearts of cruel men: "The Creator if he doesn't like your opinions shall leave you screaming in agonising horrifying stinking frying fleshly pain forever!" Although how flesh lasts longer that the time it takes to turn to ashes in fire is another religious mystery! The average length of an adult cremation at 850 degrees centigrade is generally under two hours! After that there are 113 chemical elements in those ashes and 113 known chemical elements in 7 lbs of clay proving the scientific fact of the biblical statement in Genesis . . ."Ashes to ashes—dust to dust . . . you were made from clay and you shall surely return to clay." The first man blamed the first woman, she blamed a reptile, nobody apologised hence history's horrific legacy of violence, war, crime, heartache, sickness, confusion, fear, loss, pain and the graveyard. If there is a God he surely would not have made our solar system and earth teeming with life for nothing! The universe proclaims purpose; our intellect whispers purpose, our planet declares purpose.

Richard Dawkins recently spoke to a man older than himself, Stephan Weinberg, Scientist and Nobel Prize Winner in Physics in 1979: (1933- present) in a conversation recorded, filmed and posted on his website. This well educated man wisely said to Richard: *"The greatest tragedy for mankind is that we grow old and die"* and there it is in a nutshell! The true Mystery of Mysteries. Why do we die? Every human being, either vocally or on a subconscious level dread's being old, feeble and helpless. Shakespeare said: "Some people are born old, some never grow old" Oscar Wilde said: "The tragedy of the old is that they are young" and my mother, Josie said: "When I look in the mirror and see wrinkles on my face, I know there are none on my heart."

Charles Darwin could not accept:

A God of Mysteries (i.e. The Trinity)

A creator who burned people forever.

The pagan fiction of the soul's immortality.

The prejudicial idea of one tribe being chosen.

Animal sacrifice for sins.

The idea of a murdered Jewish Rabbi—being God. (He wasn't!)

The so called divine inspiration of 66 books—The Bible.

Silly and demeaning theological arguments.

The supercilious clergy of Christendom.

The pretension that prayers are currently answered.

The fiction that real miracles take place today.

The notion of life after death.

He got most of them right and its irrelevant which two he got wrong! This was part of his tormented religious mindset so it was perfectly understandable with his vast taxonomical abilities that one day the influence of his father and grandfather should influence him. (even the great Freud would acknowledge this!) A sincere deduction from the honesty of the human heart that no God exists is sometimes perfectly understandable. Some people have suffered so much, witnessed or experienced such human brutality and misery that they cannot conceive of a loving God allowing such catastrophic pain and suffering. And in a perverse way their honest Atheism is an indirect compliment to God!

The Picture below
"THE GREATEST DECEIT IN THE HISTORY OF SCIENCE."

The root at the paradox of evil, Epicurus (341–270 B.C.) tried to sum up in his clever courtroom tactic in relation to God's perceived personality!

Is he willing to prevent evil but not able? Then he is impotent.

Is he able but not willing? Then he is malevolent.

Is he both able and willing? Whence then is evil?

However Epicurus (341—270 B.C.E.—Greek philosopher) honest questions are a credit to God in his attempt to understand the problem of pain and the terrible sad injustice we regularly hear and witness! The only sensible answer that has the ring of truth and seed of hope is in ancient Jewish and Christian documents.

The moment some people read what I have just said unfortunate assumptions take place. Some may believe I want to convert people to a particular point of view! I answer categorically I most assuredly do not! I actually do not care what people think because we are all entitled to believe anything we want. I simply know some fictions are destructive of mental stability, hence my book.

There is much common sense in parts of the Bible and the seven questions previously asked are realistically answered. Once evolution is consigned to the realm of fantasy and Darwin's debatable delusional conclusions (1000 suppositions by page 200) are exposed as a gigantic fairy tale then reason might prevail and other quotes like the following will be welcomed. Physicist and mathematician Dr. Wolfgang Smith who sees honesty beginning to prevail: said in *Origins Answer Book*, page 107 by Paul S. Taylor (1989) "Clearly, the public has been tragically misled about the scientific support for evolution, including the components of random mutation and natural selection. Dr I.L. Cohen calls it all a false theory: Micro mutations do occur, but the theory that these alone can account for evolutionary change is either falsified or else it is an un-falsifiable, hence metaphysical theory. I suppose that nobody will deny that it is a great misfortune if an entire branch of science becomes addicted to a false theory. But this is what happens in biology. I believe that one day the Darwinian myth will be ranked as the greatest deceit in the history of science. When this happens, many people will pose the question: How did this ever happen?" Dr. I. L Cohen wrote; *Darwin was Wrong a Study in Probabilities,* published 1984.

Furthermore, Dr. Wolfgang Smith, himself a physicist and mathematician says; "A growing number of respectable scientists are defecting from the evolutionist camp . . . for the most part these "experts" have abandoned Darwinism, not on the basis of religious faith or biblical persuasions, but on strictly scientific grounds, and in some instances, regretfully."

Therefore, if a book reveals why we exist in an astonishing universe, living on a planet supplied with a magnificent variety of delicious food and vital oxygen, is it not in our interest to, at least, listen and evaluate what is said against many other opinions we may have heard or inherited?

Every single baby born has no convictions! The only things it hungers for are food, love and comfort. The religious, evolutionary and political pollutions occur later and the chaotic consequences of traditions and fictions are all around us. Sometimes the more sincere an idea the more fury and anger explode when it is challenged.

Reading through Darwin's delusional dreary theory, if one cares to open at page 300–301 in his fantasy chapter called "Geographical Distribution" one can count 16 (yes sixteen) suppositions. His laughable language on these two pages, breathe and proclaim the certitude of ignorance. Nevertheless in all other scientific disciplines we demand experimental even handed evidence except one, evolution, not even realising that such acceptance is intellectual de-facto, defection and defecation!

The 16 suppositions are as follows;

The belief

Probably

Perhaps

Might have

If

Possible

Might have (twice)

Perhaps (twice)

Seem

Must have

I believe

I suppose

I believe (twice)

Reason to believe

May have

Probably (twice)

Such obvious confusion is a recorded testament to the mans horror of reality.

Now I use Darwin's literary device (suppositions) and ask you the reader to imagine, you are the patient in this analogous situation. You have been diagnosed with operable cancer and are sitting in front of the surgeon:

Patient: "Doctor is it serious?"

Surgeon: "*I believe* it's possible based on *our theory we might* have *a chance* to catch it in time *if* we are lucky"

Patient: "Am I going to die?"

Surgeon: "The current *belief* on this important question is based on the *probability* of educated *assumptions* that, *perhaps*, the pool of scientific evidence when

	collated with past mistakes and current progress *might* give us reason to believe *you may be* one of the survivors who is *possibly* fit enough to *hopefully* survive!"
Patient:	"Oh my God, is that serious?"
Surgeon:	"Not really, as *theoretically* there should be no cancer at all! However *possibly* there may be *a chance* you *might* be one of the lucky ones *if* our prognosis proves correct"
Patient:	"Does that mean you have to operate?"
Surgeon:	"Due to the prevalence of diagnostic dissembling we *assume* our analysis *may need* further opinions so we have 100 other eminent surgeons in our society we respect and you are encouraged to see any number of them you *might* be able to afford"
Patient:	"Can you give me any hope?"
Surgeon:	"Certainly, I am *inclined to believe* that *possibly conjecture* at this moment *may seem* in all *probability* to perhaps lack credibility. Therefore I am going to recommend you for further similar erudite *opinions* from other eminent surgeons."

Would you trust that surgeon to give an honest appraisal based on logic, reason, evidence, experience, empathy and empirical evaluation?

This is the language on virtually every single page of Darwin's depository of descriptive desultory delusion. The great tragedy of this book is the conclusions drawn by so many i.e., if there is no Cognitive Artistic Genetic Engineer. (God) then I am the result of blind chance no different from a lump of rock, fish, bird or animal and consequently with no hope, no purpose and certainly no future!

There are many kind moral atheists, agnostics and humanists throughout the world who would never dream of hurting their fellowman. They are not the type whose hearts are poisoned by the toxic notion that we are only a higher form of animal. No, it's the amoral violent minority who instinctively believe greed, violence, war, plunder, rape and mayhem, are simply expressions of a 'superior species' taking what rightfully belongs to the strongest and fittest of the particular tribe they perceive themselves as belonging to.

Mercy appears not to be a natural characteristic in the animal kingdom and Darwin on page 276 of his hoax quotes certain palaeontologists as recognising that:

"The inhabitants of each successive period in the world's history have beaten their predecessors in the race for life, and are, in so far, higher in the scale of nature; and this may account for that vague yet ill defined sentiment, felt by many palaeontologists, that organisation on the whole has progressed."

Similar statements from influential scientists gives the intellectual imprimatur to human savages devoid of merciful feelings for others to implement their orgies of death and destruction so their "superior species" might grow and prosper at the expense of a "weaker variety!" All the quibbling in the world shall not disprove the current and historical reality of sickening consequences in adopting Darwin's disastrous deluge of science fiction on the godless mindset of the godless blind-set. The Sunday Times Magazine, 8th, November, 2009, called "Children of the Evolution," established a connection between students who murdered classmates and acceptance of Darwin's theory.

Darwin's idea—premise—theory, "Adaption by descent through modification" has today been revealed on a biochemical, biological and biogenetic level as absolutely fraudulent. If one reads *Origin of Species* to page 328, there are 1372 suppositions and his imaginary notion of descent is regularly touched upon. However, "descent" as he wishes it to be understood through "modification" is on a biochemical level a scientific impossibility. He was ignorant of the molecular astonishing complexity of an animal or human cell—hence his phantasmagorical idea.

Darwin's drivel is a deadly poison that cripples reason, destroys appreciation, ruins simplicity, damages the psyche, commandeers common sense, devastates hope and causes havoc among the community of nations. Once natural selection is embraced natural affection is diminished.

Now to directly address Darwin's debatable delusional conclusions in his monumental and influential farcical opus. First the title:'*Origin of Species,*' I admire and freely acquiesce to his vast taxonomical abilities, however he constantly reveals his ignorance of "Origins" and *not even once* either on plants, birds, insects, fish or mammals cogently elucidates in a convincing way "the mystery of mysteries." His favourite term for this biochemical impossibility is "natural selection" and he always commences with:

An environment.

An atmosphere / oxygen.

Readily available food.

Other mysterious life forms.

An existing earth.

Water.

Males for females and visa versa.

I could give hundreds of scientific facts to prove the law of bio-genesis (life only comes from life and non-living matter never produces living matter) but that I will not do because there are copious numbers of books available which adequately undertake that and conclusively obliterate the hoax of evolution. My speciality, **Ontogeny,** the mysterious, majestic, magnificent modus-operandi of one human sperm and one egg uniting in the prophase of mitosis, to become a zygote, which contains all the genetic material needed for foetal growth. It develops into a morula and is a cluster of perfectly formed cells. Next appears the blastocyst which is now ready to become ensconced in its new home the endometrium, or uterine lining of the welcoming womb. The blastocyst becomes the embryo in approximately 8 weeks which soon becomes the foetus at around 10 weeks with every single organ of its body now formed except eyelashes and nails which occur in the eighth month. (this prevents baby from scratching itself or getting an eyelash in its eye. . . . How astonishing is this chemical time clock!)

People who have accepted the idea of evolution, are unconscious and incognizant of the scientific, biochemical, biological facts which expose evolution on a cellular level. All any person needs do is slowly read, absorb and meditate on the scientific facts presented in the brilliant book by Michael J. Behe, *Darwin's Black Box* The 1% of Darwin's disconnected disciples who accept his theory and may have read *Origin of Species* might notice on page 361 (before his recapitulation of the difficulties on the theory and conclusion) there are 1515 suppositions. However it is his very honest admission which I truly admire him for on this page. Here he calls his 1515 suppositions "facts" and unambiguously proclaims that his idea, even if unsupported by other facts or arguments, he would without hesitation, *still adopt this view.*

<u>That is like people proclaiming</u>:

The emperor's new clothes are beautiful.

"Yes there is a Santa Claus"

Muslims emphatically stating, "We know the Koran is from God"

Christians expressing the view, "The Muslims are infidels, its our book that is true"

Alchemists claiming they could transmute base metals into gold.

A naturalist gardener proclaiming "I have seen fairies in my garden."

The Earth is flat.

However listening to arguments relating to agricultural, horticultural, animal and human beings struggling to survive, is a universe of difference as to knowing how they arrive.

The problem in identifying intellectually with the scientific reality of creation is that we are corralled in with fundamentalist creationists who are embarrassing in their word for word literal interpretation of ancient scrolls. I wish it here and now to be revealed that I view organised religion as a pox on humanity and holy books in general the trigger and ignition of the disease.

I have read many so called "holy writings" and much of what is in them is laughable to the scientific modern world we inhabit. That is why, particularly among educated people world wide, fear, threats and rewards are viewed at best a morally corrupting influence and at worst a complete waste of time. The one true exception is the ancient Hebrew scrolls, confirmed as the Septuagint which translated Jewish sacred writings. Ptolemy (285—246 B.C.E) commanded 70 scholars to translate the Hebrew scrolls into Greek as many Jews were scattered throughout the Greek speaking world of that time. We are told they worked independently of one another and when each translation was collected they precisely agreed on every word and, according to Jewish scribes there are 815,140 Hebrew letters in their accepted thirty nine books.

I understand Darwin's confusion with the theological constructs of his day. I commiserate with his agony on the death of his beloved mother as a young boy and his perplexed innocent mind trying to come to terms with the objectionable, obnoxious, abstruse, absurd teaching that "a loving God took her!" This was the seed of understandable destructive enmity against God which grew in him as years progressed.

I admire Darwin's father, Robert, who was a closet freethinker and his grandfather Erasmus who wrote *Zoonomia* which included many evolutionary concepts and both were, I am delighted to note, "anti-religion." . As a child, Darwin appeared to retreat into a world of fantasy. For example he wrote in his autobiography, "I may here also confess that as a little boy I was much given to inventing deliberate falsehoods, and this was always done for the sake of causing excitement." Desmond and Moore (pro evolution authors) published: *Darwin—The life of a Tormented Evolutionist* in 1992, and wrote, "Two of the most influential people in Darwin's early life and thoughts were his father, Robert, and indirectly, his famous grandfather Erasmus. Although Erasmus died before Charles was born, his father made sure he was familiar with his grandfather's writings on evolution. Erasmus Darwin included many evolutionary concepts Charles would later adopt, and had been a successful physician, as was his son, Robert, and both were decidedly anti-Christian although careful to disguise their ideas in public.

The name of Darwin was already associated with subversive atheism. Dr. Robert was himself a closet freethinker . . .

Charles Darwin eventually rejected Christianity, in part because he could not accept the fate he understood it to decree for unbelievers such as his grandfather, father, older brother and even himself. He wrote in his autobiography: "Thus disbelief crept over me at a very slow rate, but was at last complete. The rate was so slow that I felt no distress and have never since doubted even for a single second that my conclusion was correct. I can indeed hardly see how anyone ought to wish Christianity to be true; for if so the plain language of the text seems to show that the men who do not believe, and this would include my father, brother and almost all of my best friends, will be everlastingly punished, and this is a damnable doctrine."

Darwin was an attention seeker and wanted praise. He would do anything in school for the pure pleasure of exciting attention and surprise and he cultivated lies which gave him pleasure like a tragedy. He told tall tales about natural history and he once invented an elaborate story designed to show how fond he was of telling the truth. It was a boy's way of manipulating the world. He often told lies about seeing rare birds and the lies were not connected to any sense of shame, more accurately, they mirrored a search for attention. He wanted to be admired—lies—and the thrills derived from lies—were for him indistinguishable from the delights of natural history.

I agree with his conclusion after reading Charles Lyell's book that the earth is quite likely millions of years old (and the universe possibly billions!) But so what!

Why fight and argue over when or how it began, when in fact we can never truly be certain! The real question is WHY???

I understand as a naturalist his sailing into the Galapagos many beautiful islands for five weeks and being perplexed by the variety of beautiful birds, lizards, plants, tortoises (he once drank the urine of a large tortoise to see if it was potable..... it was!) fish and in particular finches and their different beak shapes and sizes!

On arriving back in England after an arduous five years travelling around the world his sailing days were over. He had read: Thomas Malthus book, *Essay on the Principle of Population* and was struck by the similarity between mans' competitive struggle for limited resources and the constant fight for survival in nature, providing a possible basis for evolution-natural selection, the survival of the fittest. Here he had at last, a theory by which to work.

7 MOTIVATIONS

He now had his concept (with no scientific evidence)

He now had the mechanism (he imagined)

He now had a subject he loved; (observing nature)

He now had a purpose (to make a name for himself)

He now became a scientist (but denied true science)

He now was a taxonomist (a very good one)

He now despised organised religion (understandably)

The reason I quote those paragraphs is to highlight his statement in reference to what he believed were his "heretical thoughts" which he kept to himself. He actually said it made him feel like he was "confessing a murder." I smiled at this but instinctively knew what he meant because of the intellectual prison of Victorian society. The hypocrisy, snobbery, arrogance, rules, customs, conventions of the ruling classes to which his wife and family subscribed (which he was a part of) and the silly superciliousness of the laughable clergy: In our open educated society today it would be like admitting that we like sex or chocolate or soccer! Nevertheless it perfectly highlights the vice-like grip various churches held their victims in at that period of world history.

Two Devastating Deaths in His Family:

Next, he received two devastating blows to his young family. According to biographer Janet Browne, the death of his beloved daughter, Annie at the age of 10, followed a year later by the death of his first born son, William, caused great bitterness towards God. Their deaths were the formal beginning of Darwin's conscious dissociation from believing in the traditional figure of God. Bleakness swept in. The gradual numbing of his religious feelings and the godless world of natural selection he was even then still creating, came implacably face to face with the emptiness of bereavement.

Yet ironically some might say Darwin was a victim of his own theory of natural selection because of the genetic dangers of inbreeding. In 1839, he married Emma, his first cousin. Both families had intermarried through first cousins for some time, a dangerous trend for heredity. Twenty-six children were born from these first cousin marriages: 19 were sterile and five died prematurely, including Darwin's daughter and first son. Some suffered mental retardation or other hereditary illnesses, as was the case with his last son. All these affects engendered great hostility toward the idea of a personal, intervening God." The religious teaching then prevalent and voiced at funerals was the disgraceful notion that a loving God took (killed!) them. The reality is God needs nor takes nor removes humans from their families.

Note: I recently went to see the movie; '*Creation*' and shed a tear when Annie died.

A Devil's Chaplain

Darwin wrestled at this time with publishing his theory, fearing ostracism, the strain showed and in a letter, Darwin blurted, "What a book a Devil's Chaplain might write on the clumsy, wasteful, blundering low and horridly cruel works of nature!" It was a book that Darwin feared he might be accused of writing, a book that would reveal him as an unbeliever and open him to punishment—like the original Devil's Chaplain, Rev. Robert Taylor, the Cambridge graduate and apostate priest who was twice imprisoned for blasphemy." (Darwin- *A Devil's Chaplain*—online edition)

To lose ones mother at the age of eight, a beloved ten year old daughter and a year later your first born son is an attack on any person's grip on sanity! Consequently he declared war on religion and forged a partnership with pseudo-science. As far as his solitudinarian existence was concerned it was profitable towards his ultimate goal—the publication of his book. I know he was sickened by nature's cruelty but I would have been sickened more by the shocking poverty and children working long dreary hours down coal mines for pennies, as well as child prostitutes selling their abused flesh for shillings on many streets in England which he and the ruling classes chose to ignore publicly while many so called gentlemen used them privately.

I do feel sorry for him in that he suffered with poor health for 40 years (evidence suggests some of it psychosomatic!)

Not only was he suffering from what seemed to be psychologically induced illnesses, but he was also racked with doubts about his own book. He confessed to some fellow scientists: "It is a mere rag of a hypothesis with as many flaws and holes as sound parts . . . A poor rag is better than nothing to carry one's fruit to market in" To another colleague he wrote, "I have devoted my life to phantasy (sic)" Quoted by Desmond and Moore, pages 475–477, *Darwin The Life of a Tormented Evolutionist*.

As Desmond and Moore explain, "Plumbing the radical depths Darwin saw the cataclysmic consequences. Once grant that species . . . may pass into each other . . . and the whole fabric totters and falls." The Creationist "fabric" and all it entailed was his target. He peered into the future and saw the old miraculous edifice collapsing" (page 243)

A Man for The Times

The age of positivism had arrived, promising science would lead to an epoch of constant scientific and material progress, ultimately answering all of man's questions and solving his problems without the help of religion. It was also a time when the churches of Britain were viewed by many radicals, like Darwin, as corrupt and outdated. (they surely were!)

Darwin proposed a theory that essentially displaced the creator God, with only physical and undirected mechanisms such as natural selection and adaption doing the creating. "His vision," state Desmond and Moore," was no longer of a world personally sustained by a patrician God, but *self generated, from echinoderms* (marine creatures such as starfish) to Englishmen, all had arisen through a lawful redistribution of living matter in response to an orderly changing geological environment."

It should be noted that in later editions in *Origin of Species*,' Darwin did add the term "Creator" in a few places and in his conclusion, in one place stating:, "There is grandeur in this view of life, with its several powers, having been originally breathed by the Creator into a few forms or into one" (this phrase "By the Creator" was inserted in his second edition). Yet he later confessed to his outraged colleagues that this impression of theistic or deistic evolution was to soothe the feelings of his Christian wife and of a like minded public. Even so, Darwin admitted to wavering views and claimed to be agnostic. In 1879 in a letter he wrote, "I have never been an Atheist in the sense of denying the existence of God . . . Agnostic would be the more correct description of my state of mind."

(Darwin to J. Fordyce, published by him in 'Aspects of Scepticism' 1883).

He was a gifted naturalist but wavered in his actual evolutionary convictions and, at times, acted the hypocrite in company of his devoted, conventional, Christian wife Emma, and he was also sometimes a coward in facing his colleagues with his true beliefs.

Consequences of his Theory

It is probably an axiom to state that religions help keep good people good through repetition of rewards for moral behaviour and punishments for errors. However, throw his metaphysical concept into the ring that there is no First Cause—no God, no *Creator* because pseudoscience tells us, then his dead idea invented by an understandably spiritually scarred man called, "natural selection" or transmutation or evolution and you have a license for earth-wide deviancy because someone has linked, likened us to animals in our struggles throughout our lives.

Men like Richard Dawkins, Christopher Hitchens, Daniel C. Dennet and Sam Harris are evidently moral, civilised and generally gentle in their dealings toward their fellowman. However, they must surely (surely) recognise that there are men on this planet who cannot be controlled by religious offerings or lawful constraints. Once they are influenced by evolution to accept that we are only a higher form of beast, then the mask of civilisation drops. There is no stopping the savagery they would engage in to conquer "inferior species" and secure their goals of establishing the, "survival of the fittest!" That's a fact. . . . arguing to the contrary is dishonest, hypocritical, pointless and illusory.

The results of Darwin's theory of evolution were dramatic. Atheism and secularism became widely popular. As one of today's most ardent modern supporters of Darwin and Atheism, Richard Dawkin's has famously said, "Darwin made it possible to become an intellectually fulfilled Atheist" (*The Blind Watchmaker*, 1986, page 6)

Scientific materialism spread like wildfire. Karl Marx, the father of communism, out of gratitude to Darwin, sent him *Das Capital* his principal book on communism. 'Although developed in the crude English fashion,' he wrote to his communist colleague Freidrich Engels, "this (Darwin's *Origins of Species*) is the book which in the field of natural history, provides the basis for our views." To another he wrote that Darwin's work "suits my purpose in that it provides a basis in natural science for the historical class struggle."

This evolutionary backing eventually helped establish the philosophical framework for the twin scourges of communism and atheism in Russia, China and many other nations. As Darwin's ideas gained respectability, moral absolutes were increasingly questioned. If there is no Creator, then it seemed all things are permissible. If there is no God, then there are no ultimate consequences. If there is no greater authority than yourself, then the rules of, 'survival of the fittest' are in effect and support the idea that you can succeed by any means in applying the law of the jungle—only the strong survive. To cap it off, Darwin wrote in 1871 his, *Descent of Man*, describing human descent from apes, a book with considerable baseless speculation and even racist claims—including that of white supremacy (as whites were reckoned as further than blacks from apes along the evolutionary advancement chain).

Hitler later used some of those ideas, called "Social Darwinism" in World War II to eradicate millions of Jews and others he thought racially inferior. He said: "Nature is cruel, therefore, we too may be cruel . . . I have the right to remove millions of an inferior race that breeds like vermin! . . . natural instincts bid all living beings not merely conquer their enemies, but also destroy them" (quoted by Hermann Rauschning, *The Voice of Destruction*, 1940, pp. 137–138)

In effect, Hitler could say he was applying the theory of evolution and only quickening the inevitable end of the weak. This was necessary to make room for a fitter, superior race. It gave him what he thought was a scientific and moral validity for his twisted obsessions–and some 55 million people died in World War II largely because of these warped views.

Once again I accuse most religious leaders in the world of complicity with all the inhuman wars of the 20th Century. Stop pretending peace is your policy because, when war breaks out your sympathies lie with the leaders of your particular tribe.

No Evidence

"What can be asserted without evidence can also be dismissed without evidence," cleverly wrote the able writer Christopher Hitchins in his understandable defence of an imaginary idea of evolution. Most people who believe it begin their support thus: "Scientists have proved such and such." I have listened to this for 40 years: I have always offered my name and address to receive proof in the mail and I have waited 40 years! They invariably forget that scientists, geologists, zoologists, anthropologists, palaeontologists, naturalists, teachers, and others are mere human beings with feelings, prejudices and superstitions like most people. The upper echelons of academia are like gorillas that beat their chests and holler the loudest to frighten off enemies and attract females. These bright (Brights!) individuals subscribe to a culture where adherents to this supposedly proven fiction are in the inner sanctum where the status quo is venerated.

Sigmund Freud had his inner circle in Vienna and one of his students, Alfred Adler, rejected his exclusive emphasis upon sexuality as our unconscious motivator. But what if he was/is right? Power, money, fame are powerful aphrodisiacs for weak or powerful women and who is not intoxicated by success? The current talent shows on TV worldwide demonstrate the grip popular culture has on the psyche and the ends (plus humiliation) people are willing to endure for their 15 minutes of fame. The educated promoters of the theory of evolution are no different. They are, like me, susceptible to flattery, compliments and advances of the fairer sex.

No Connection

I wonder if a well known colleague got up on stage at one of their scientific conferences and said the following:

"Ladies and gentlemen I have an admission to make to you today: after believing the theory of evolution for 35 years, last year I accepted a challenge from my son who is a medical doctor, to examine ONTOGENY! As you may know this is the study of one successful male sperm out of around 500 million uniting with the female egg and fusing to produce the golden prize, a new life. This study alone of the zygote governed by chemical laws and beautiful mathematics is under microscopic analysis, mesmerising. I followed the cellular division for some days until I observed the blastocyst which I exhibit for observation. The scientific fact is that all over the world right now from an act of beautiful pleasure, a child of love shall be born from the astonishing division of one unbelievably complex cell into two. Stop and think. Think of the precise molecular combinations unseen at architectural construction on a biochemical level to produce four cells and so on. I carefully and open-mindedly followed the progress every day on a number of scientific and physical levels in the maternity hospital where my son works and have been staggered at my physiological ignorance of the process from conception to birth. I had no religion then and I have no religion now, because as a humanist I see organised religion as a diseased scab on society. However, what has changed is me. Last year I casually accepted the theory of evolution and rejected the idea of God.

Today I tell you as a scientist that for years I, with many of you, accepted this theory as true and joked, how stupid some of the fundamentalist Christian creationists are and possibly still are in their six day origin theory when science apparently establishes on a number of levels the earth is much older! I know how sincere you are and I wish to assure you of my respect for your ongoing work in so many scientific disciplines.

However, I would rather be sincerely wrong than insincerely right. I must have freedom to assert my conviction that after studying the mesmerizing staggeringly complex interconnected cell, one cell, to billions of perfect cells in my son's baby. I have drawn the conclusion based on pregnancy and ONTOGENY alone, that there is a master genetic architect our society labels God and which I now acknowledge as *The First Cause.*

Do not ask me who he or she is after my surprising little speech today for I admit I do not know! However what I do know is that I am a scientist and my conclusions are in perfect harmony with the dictionary definition of science. I am not a candidate for conversion to any, and I mean any, religious club. I love my freedom to think, examine, tabulate, categorise, and conclude based on what I believe to be facts. I stand before you today confident that each and every one of you respects my right to choose what I decided to accept and hopefully you will continue to accept me. Thank you!"

I wonder and I would like to believe this new construct based on what he examined, studied and concluded on the balance of probability, YES there is a supreme cognitive architect—first cause, would be welcomed among his colleagues!!!!!

One Simple Question

Since I only deal in demonstrable facts, I now wish to give an example to illustrate mans' KNOWN, as opposed to imagined, tenure on this planet. To begin, I list seven major world empires.

United States of America.

Great Britain

Rome

Greece

Persia

Egypt

Babylon

Before the Egyptian and Babylonian cultures there were other cruel, violent dynasties all scrambling, manoeuvring, warring, scheming and killing for political power and as always their murderous intent sanctified by hired paid ignorant priests . . .

Also, since I detest evolution for the hoax and fiction and unscientific hogwash it truly is, I now ask you to visualise a group of seven individuals sitting at a round table and each has agreed to answer YES or NO to a simple question:

DO YOU BELIEVE YOU CAME FROM A WOMAN WITH THE HELP OF A MAN?

Each time this logical question is asked all have agreed the mean average throughout recorded history for a generation is twenty five years.

This exercise may appear simple, however Albert Einstein said; "Without simplicity there is no greatness." He also remarked "To see life through the eyes of a child is to remain perpetually young." But I prefer Oscar Wilde's analysis of why adults reject powerful obvious logical deductions of elementary propositions. In his play *Lord Arthur Savile's Crime*, it is said; "To Lord Arthur it came early in life—before his nature had been spoiled by the calculating cynicism of middle age." Or who has not heard of the famous story told by Hans Christian Anderson, "The emperor and his invisible clothes!" In this moral tale many of the people and all of his court are telling him how magnificent his invisible clothes were! Emboldened by his vanity and their pretentious nonsense he decides to go public; everything is going according to plan until a little boy shouts, "Why is the king naked?" The spell instantly broke and reality immediately surfaced as the old adage once again proved correct; "You can fool some of the people some of the time, but you cannot fool all of the people all of the time!" It took a child's natural innocence and devastating **logic to expose A RIDICULOUS IDEA who everyone including the King** *knew in their hearts was a FICTION.*

I come now to the core of the matter illustrated by a round table with seven educated individuals seated and waiting for the process to commence. They are:

Scientist— Woman

Historian—Man

Mathematician—Woman

Paediatrician—Woman

Evolutionist—Man

Naturalist—Man

Philosopher—Woman

All have agreed the average generation is twenty five years during the known major seven world empires and before.

Each agreed to answer YES or NO to the only question allowed in this pure scientific historical factual undertaking; "Do you believe you were born from a woman with the help of a man?" That woman and man being your mother and father ...

Every person present male and female were forced by connecting threads of logic, evidence, experience, self respect, sanity, common sense and reason to unanimously state in answer to the simple question, YES! We have now moved back in time twenty five years to 1984. The same question is now proposed to each of the seven people in relation to their grandparents "Do you believe that woman (your mother) was born from a woman with the help of a man?" and again the seven threads of evidence produced, YES. We are now back fifty years to 1959. Same question for great grandparents produced the same scientific, historical, factual response YES which brings us back to 1934 . . . again the same question elicits the same sane response YES which brings us back in time 100 years to 1909 . . . That is not a great deal of time, however four generations have come from the first father and mother in this exercise commencing in 2009.

Rather than laboriously and repetitively go through this formula twenty five years at a time the participants have all agreed to move through the exercise 100 years or four generations. Again the question: "Do you believe the next four generations came from women with the help of men? YES or NO." Each person unequivocally responded, YES which brings us back to 1809 being a total of eight generations or 200 years. Again same question, same response and now back to 1709 or 300 years and 12 generations; same question same response and back to 1609 which is 400 years and 16 generations and finally to bring us back 500 years to 1509 and twenty generations the same question was asked with the same YES answers. The historian is now asked to demonstrate on the world map and point out the cities that have all but disappeared as we understand modern cities. There are on this map many with small light bulbs and the following seven disappear.

Los Angeles	*USA*
New York	*USA*
Sydney	*Australia*
Hong Kong	*China*
Montreal	*Canada*
Johannesburg	*South Africa*
Auckland	*New Zealand*

We have only gone back to 1509 which is 500 years or twenty generations, but the world as we know it today bears no resemblance to the world then! Our panel for expediency have agreed to imagine we have consequently counted another 500 years and all have agreed YES which brings us back to 1009 A.D. and forty generations. Each person in the experiment acknowledges by moving back in time at twenty five year intervals everyone has had to say YES and particularly after hearing the paediatrician inform all that for pregnancy to occur there must be present in perfect synchronisation on a biochemical level the following seven vital ingredients at any and all times throughout medical history:

A perfect male sperm.

A perfect female egg.

A perfect penis.

A perfect vagina.

A perfect fallopian tube.

A perfect womb.

A healthy male and female body.

Anything less, sometimes even the tiniest flaw and pregnancy cannot take place as many couples have sadly discovered, nevertheless it's a miracle if it does and a miracle if it doesn't.

All have now agreed to go further back another 1000 years to 9 A.D. which is 80 generations. Our historian is now asked to remove a further seven well known cities and the lights go out on:

London

Dublin

Berlin

Paris

Moscow

Madrid

Prague

Each person realising that 25 years is a relatively short time on a physical level has wisely not interrupted for the past 2000 years and answered YES to the question: "Do you believe she/he came from a woman with the help of a man?" Fourteen well known cities have disappeared and we are only one third through known recorded history. Again the question for another 1000 years! All are in agreement and say YES. Our mathematician is asked to forget the nine years and go back 1000 years; she agrees and gives us the date 1000 B.C which is 3000 years ago (approx) and 120 generations. This is 50% through recorded history and it is not difficult for the human mind to grasp known inhabited areas of our planet shrinking and our grandparents of 120 generations ago may have lived in Northern Europe or around the Mediterranean! (Who knows for sure?)

The next 1000 years all are in agreement and say YES . . . that is 4000 years ago and 160 generations. More lights go out over more cities in the following countries.

Mexico

Brazil

Nigeria

China

Japan

Korea

Greenland

The early Bronze Age had likely begun and some of the earliest forms of known writing were cuneiform on clay tablets in Sumeria and hieroglyphics in Egypt. The great pyramids were possibly still under construction. The Babylonian culture and commerce are predominant in the Mediterranean region of western Asia and the first King of Egypt was long dead.

Finally the seven participants are asked will they accept 2000 years in one giant leap back for mankind. All agree rather than count back 80 more generations. At this prehistoric period an extraordinary number of facts emerge that are historically indisputable, and this period of human misery/history is 6000 years approximately.

RESULTS

Not one known city remains.   *- FACT*

No humans are on earth.   *- FACT*

No writings of any kind are older than 6000 years.
                                           - FACT

Animals—fish and birds are plentiful.   *- FACT*

There are no holy books.   *- FACT*

Earth at peace.   *- FACT*

Man and woman are about to arrive, ALIVE.
(as the fossil record proves)   *-FACT*

The purpose of this factual section of my book is sevenfold;

To highlight scientific facts.

To elucidate historical facts.

To highlight the majesty of pregnancy.

To demystify the secrets of ONTOGENY.

To point towards a First Cause.

To expose, no cause / chance / random design, as scientifically FRADULENT.

To make evolutionists—stop, think, and evaluate known evidence as opposed to speculative specious meanderings.

If the above is true, it states simply and powerfully that woman and man dramatically and quietly appeared suddenly! It therefore means all the erudite suppositions on so called origins are simply SPECULATIONS. I am not directing anyone to any religion or holy book in particular as I reject all religions as nothing more than mans inner longing for meaning at best and worst a curse on humanity.

As far as so called holy books are concerned there is not that much *actual holiness* in them. However one in particular I find very interesting, a Jewish scroll called Genesis! It was written about 3500 years ago, possibly by Moses. Genesis is the Greek word for creation and in the absence of intelligent alternatives makes a lot of scientific and common sense. The following seven reasons form the core of hope for a bright future as gleaned from that document.

Humans were created for eternal youth.

Man was not made to die.

Gods real name is YHWH (or Yahweh—Jehovah)

A family were chosen to reflect Gods purpose (Jews)

A messiah would win back what the first man lost.

Our loneliness, pain, short life and confusion are temporary.

God's government on earth will restore peace and eternal youth, one day.

Six thousand years ago at 25 years per cycle, we count 240 generations back to our *provable origins!* Some 6000 years in the history of the universe is but a blink! Living as we do only 70 years on average is ludicrous! (a tree can live far longer) Religions do not have the answer to mans sad predicament and if some of them appear to offer truth, common sense and hope, the price is usually conformity and intellectual suicide!

STOP READING NOW:

Think.

Reason.

Imagine.

Wonder.

Reflect.

Realise.

Question.

I often sit alone in my little back garden on quiet evenings, look up at the sky and say as Sean O' Casey, the Irish playwright, wrote in his famous play, *The Plough and the Stars*, "O the stars, the stars what is the stars?" and my intelligence tells me; the stars the stars is the mirror of God's garden reflecting down the spark and twinkles of his unfathomable creative wisdom as they whisper to me: " Appreciate, John, and enjoy—enjoy, the stars—the stars."

6,000 YEARS AGO: IMAGINE.

A world without people!	-FACT
An earth without cities!	-FACT
A planet teeming with life!	-FACT
An earth free from graveyards.	-FACT
A virtual paradise, awaiting humans.	-FACT
An earth in harmonious unity.	-FACT
A magnificent result of a supremely intelligent cause.	-FACT

What I have written are facts as opposed to speculation. A mere 240 generations ago is around 6000 years. A universe that silently speaks, mesmeric wonder. A milky way that proclaims mathematical genius. An earth that sings to life with living creatures in the oceans, air and beautiful lands. But something was missing!!!

I write as a man cognitively convinced that nothing comes from nothing and our Universe, Milky Way, Earth and all life forms in their astonishing and interconnected interactions are the intelligent effect of an effective knowing cause. Once a Cognitive—Artistic—Genetic—Engineer (which unfortunately in most peoples daily vernacular is called not by a name but by a title: GOD) is focused on and the penny drops, yes! There is a SUPREME ARCHITECT—everything changes . . .

Atheists accept the graveyard as their fatalistic future. Religionists believe in various fantasies based on devotional delusional FICTIONS founded on inscrutable mythical writings written mostly by delirious deluded demagogues.

Basic common sense in the mind is the heart's twin in convincing us that when a baby, adult or old person dies, they are . . . DEAD. . . . DEAD . . . DEAD . . . NOT LIVING. NOT

ALIVE . . . GONE; about to be cremated or buried in the silent dark deep cold ground! That is a fact. We who are healthy, relatively young and loved are happy with no desire to perish. Yes, oh yes, we want to continue enjoying the pleasures of daily living with the thoughts of more in the future with our lovers, family, friends and partners . . .

Yet on 11 September 2001 some 3000 innocent, responsible human beings had their flesh, breath, hopes and dreams savagely shattered, sheared, crushed burned and buried in the single largest graveyard filled in one day in the indifferent smoking ground of New York City. The people responsible? Muslim miserable, misologistic, misogynistic, misoneistic, misanthropic, misfits! And why? Because of perceived past historic wrongs, modern technology and a malignant, malicious malfeasance born from seed sown in a so called holy book, the Koran. With or without religion good people will do good things and bad people bad vicious cruel acts. However once the toxic poison of deeply sincere religious belief infects once pure and innocent minds there is no cruelty conceived, by hearts believed, that can stop their fellowman bleed—bleed—bleed!

The ingredients of that life and soul destroying poison are;

Rock solid conviction.

Sensual rewards.

Indifference to suffering.

Hatred of humans.

Belief in a book.

"God's will."

Brainwashing.

Once they are in place no horror is too horrible to injure and inflict on "Gods enemies . . ."

Then when sudden savage catastrophic violent death and destruction rain down on men, women and children, many survivors curse God and lose religious faith, others lose their reason while some remain cool and stoical.

A few (very few!) appear to understand childhood influences and the power they hold over us virtually everyday of our lives and accept the uncertainties, unfairness and oftentimes randomness of a vicious volatile violent world.

So now the question should be asked again, "Where do we go when we die?"

SEVEN ANSWERS

Good Christians -   Heaven

Bad Christians -   Hell

Good Muslims -   Paradise

Bad Muslims -   Hell

Some Jews —   The Grave

Some Hindus—   Reincarnation

Atheists—   Non existence

As a lover of science reason reality and common sense, I agree with atheists and educated Jews: The dead are dead and gone forever if there is no God! On the other hand if there is then surely HE is capable of bringing back the dead to life! However that is a subject fraught with tension which lends itself to human anxiety and conflict in conversation. This chapter is about exposing the fictitious nonsense of evolution and once that is scientifically established then each person can decide for themselves if such a future outcome for the dead is a real possibility. If the SUPREME ARCHITECT does not communicate with mankind then how are we to be free of our pathetic questions and ignorant answers? The wisest people, greatest thinkers, teachers, philosophers, leaders, priests and poets have all disagreed on mans' purpose in what appears to be a cold indifferent universe and our ultimate destiny! Millions throughout mans' painful history have gone to their graves, brutalised as farming slaves, ignorant and confused about life's meaning or purpose . . .

Why in Anglo-Saxon parts of the world, in particular, has this non-provable fiction, evolution, based on so little, come to be accepted by so many?

SEVEN REASONS

Confusion regarding (God's) ultimate purpose.

Ignorance of true scientific facts.

Peer pressure.

Antipathy towards Deism—Theism.

Injustice, pain, and random death.

Ridiculous religious teachings.

A book called Origin of Species.

This last mentioned speculative tome should have been called "Observation of Species" because the one thing it never establishes is actual origins! Page 42, paragraph 2 honestly admits the following: "I think these views further explain what has sometimes been noticed—namely, that *we know nothing about the origin* or history of our domestic breeds."

There are 384 pages in this elaborate fantasy of 1515 suppositions. In a six-page introduction on the very first paragraph of the first page Darwin mentions "Origin of Species" for the first time and quotes the British astronomer, Sir John Herschel (1792–1871) in connection with his books title as to origins; and calls them; "That Mystery of Mysteries." There are five suppositions on page one alone:

Might

Possibly

Speculate

Seemed to me

Probable

Page 12, paragraph 1 states; "No doubt errors will have crept in, though I hope I have always been cautious in trusting to good authorities alone." Remember those "good authorities" rarely agreed with each others opinions and none of them had access to modern microscopes which today reveal on a DNA and biochemical level the impossibility of slowly accruing mutations (which are mostly destructive) changing one particular variety into an identifiable new species. Note: Most of the DNA *must be present* for a cell to function on a chemical level. For example, if a magnificent new automobile after years' of design and assembly has no spark plugs it cannot function as intended. This means certain facts must be faced;

SEVEN FACTS

A time existed when there were no species.

Species suddenly appeared. (as fossil records prove)

None mutated into a different kind.

Variety is a subordinate of original species.

Natural selection like "spontaneous generation" is, was, and for all time shall remain A FICTION.

Natural selection is scientifically, anatomically, genetically and biochemically IMPOSSIBLE.

Spontaneous generation (the idea that microbes/life appears form nothing) was demolished by Louis Pasteur's scientific experiments in the 19th century.

Darwin's convoluted concoction has been exposed as an idea that was never based on science but founded on speculative fantasy and observational data accumulated over a lifetime.

Charles was a good man with a very bad idea driven with conviction that we evolved from animals and they came from reptiles and they came from fish and they sprang from slime in the sea and mud and it just arrived from nothing. I kid you not. The sad and obvious fact is reason was rejected for a fantasy he perused and pursued most of his painful life. This delusion fed his anxieties and with his vast taxonomical and observational skills tried to put square pegs into round holes.

Time is the trick he and his disciples use as a smokescreen for untenable un-provable, unbelievable, unscientific theories. Time is destructive NOT constructive. Every single bird, animal, fish, fruit, flower-food is perfect in its homomorphic, skeletal, chemical, atomic and biological structure. A half penis and quarter vagina with an undeveloped womb and a developing sperm with an unformed egg, shall never, never, never produce a healthy offspring. There are millions of healthy women and men worldwide who have minor defects in their reproductive systems and are childless. So evolutionists please stop fantasising and speculating what you imagine were our origins 50 million or a zillion years ago!

NOTE: Randomness, mutations and chance almost always lead to failure. Chance creates mistakes and problems more than 99.999 percent of the time.

Consequently evolution is:

A Fantasy

A Fallacy

A Farce

A Farrago

A Failure

A Fairytale

And a Factious Fatuous Factitious Fallacious FICTION.

Six thousand years ago there were no humans on this earth!
FACT

Seven Death Blows to the World's Most Destructive Fiction:

The inviolable law of biogenesis. -FACT

The biological science of ONTOGENY. -Mesmerising

The unselfish heart.-Unanswerable

The mathematical precision of our universe. -Unalterable

The mesmerizing majesty of a living cell. -Our Internal universe

The biochemical molecular magnificence of life. -Reasons reason

Science—Facts & "Beyond a Reasonable Doubt." -Mental Medicine

Before I elaborate on the above incontrovertible, indefeasible, indestructible, impossible to refute teleological seven propositions, I wish to make a few points on "belief" and how and why we believe the things we do! Oscar Wilde said, "To believe is very dull, to doubt is intensely engrossing, to be on the alert is to live, to be lulled into security is to die."

The question must be asked, "How did so many, adopt so little, to provide so much, that this fiction is the bedrock of delusional delirium?

Andrew Newberg, MD and Mark Robert Walkman published a book in 2006 titled, *Why We Believe What We Believe*, page 9, "The adult human brain is childlike in another way: we automatically assume that what older people tell us is true, particularly if the idea appeals to our deep seated fantasies and desires. Advertisers often take advantage of this neural tendency, and even though consumer advocates and some laws have helped to level the playing field, the general rule 'Buyer beware' still prevails. Magazine covers and full page adds promise instant beauty, fabulous sex, intimate communication in five easy steps, and we believe them, often ignoring obvious deceit."

Also "The propensity to believe that other people's values (i.e. beliefs) are misguided has fostered centuries of animosity throughout the world."

And "Neurologically such prejudice seems rooted in human nature for the human brain has a propensity to reject any belief that is not in accordance with one's own view."

And that is the paradox-view and tragedy of the unbelievable, untrue, un-provable, unimaginable dreary theory of evolution. Millions have assumed the pseudoscience which plagues our planet is true! This poisonous fiction is further entrenched by well educated polemical polymaths who invariably use the weapon of scorn to superciliously surlily suppress, courageous anti-religious, anti evolutionists who trumpet reason yes reason as our weapon of choice in demolishing that destructive soul destroying useless pointless hopeless hapless heartless historical hoax!

I hold great respect for scientific minds *when they stick to provable results* as many do on most disciplines, except one! When it comes to the fantasy of evolution coupled with the twin belief of godlessness, absurdities take the place of absolutes!

There are seven reasons why many in the scientific community accept without proof this toxic myth:

Disbelief in a scientific first cause.

Rejection of reason.

Biochemical facts ignored.

Confusing God with religion.

The stupidity of organised faiths.

Unwillingness to examine evidence.

Fear and peer pressure.

There is no law against daydreaming but scientists should not engage in it. The canard of "natural selection" as developed in *Origin of Species* cleverly buried under 1515 suppositions needs to be finally exposed as the hoax it truly is. Since science is the experimental assessment of reality and reality is the totality of real things and events (as opposed to imaginary things) then without fear let the facts speak.

The root meaning of bio (a Greek word) is life. Living organism: and the dictionary definition of Genesis is; "The coming into being of something: The Origin." This is taken from the Greek word "Genes" or Gena meaning: born or creation but called in Hebrew after its first word: Be-reshit—"in the beginning." This means in pure incontrovertible scientific language that life only comes from life and non-living matter has not, cannot and shall not produce living organisms. Now before some scientists become apoplectic and induce a paroxysm of righteous wrath, remember I too possess an Achilles heel. What this is I am not saying because I know enough about human nature to realise it most assuredly would be used against me, however, be certain about one thing it hurts no one but possibly myself!

I have no fixed opinions if evidence, reason, logic, proof and common sense on any subject is shown to me. I shall be the first to embrace and face facts. When I know nothing about a subject, I express no opinion on it. The more scientists learn about the beginnings of life the greater the wonder. For me, ONTOGENY is "The Joy of Joys," "The Miracle of Miracles" and "Wonder of Wonders . . ."

Understanding how something works is not the same as understanding how it came to be. Evolution is now a tradition or an accepted unquestioned doctrine just as surely as many discredited ludicrous laughable teachings of religions. I say now: unburden and demolish fictions that demoralise our humanity, diminish compassion, reduce morality and views human life as a meaningless meandering mammalian no better or different than a fish or piece of rock.

Teaching children evolution ("evil-ution") is a ludicrous obscenity and programmes them to rationalise good and evil as purely subjective. A new study by The Barna Group, a Christian research company in California, USA, shows young adults and liberals struggle with morality—Note the following:

"Americans have redefined what it means to do the right thing in their own lives. Researchers asked adults which behaviours they have engaged in during a one week period. Behaviours listed included: pornography, profanity, gambling, gossiping, sex outside marriage, retaliation, getting drunk and lying."

It also found that "one of the most stunning outcomes . . . was the moral pattern among adults under 25. The younger generation was more than twice as likely as all other adults to engage in behaviours considered morally inappropriate by modern standards."

For example, almost two in three in this group had used profanity publicly, almost two in five had lied or engaged in sex outside marriage, a third had viewed pornography, and one in four had gotten drunk.

Not surprisingly, adults who described themselves as liberal on socio-political issues were *twice as likely* as those who described themselves as conservative to engage in these activities. Atheists and agnostics were about *five times more likely* to participate in these behaviours than those who described themselves as evangelical Christians.

I immediately state I am not, nor ever shall be an evangelical Christian, in the modern sense as is generally understood. I respect a Jew (Yeshua / Jesus) who never saw nor read the Greek scriptures which are erroneously called "The New Testament." He, it is genealogically recorded, descended from a Semite, Abraham who did not have the Hebrew scriptures falsely called "The Old Testament" and he descended from the first man ever made whose name means "Red Earth" and he had NO RELIGION. . . . no holy books, churches, mosques or synagogues to attend. Based on those facts it appears the earth was to be free of religion and all its divisive toxic fruits. Even Richard Dawkins, the avowed atheist, admits his affection for the Church of England! Why? That's like a Nazi after World War II discovering the hideous cruelties inflicted by that poisonous political cult and exclaiming similar sycophantic nonsense about "the ceremonies and atmosphere" produced from opiated mass hysteria during that twelve year reign of fear, madness, terror, brutality, savagery, and all sanctified, accepted, tolerated and promoted by the Church of England's sister churches in Germany. Wake up Richard and smell the coffee!!!

A DICE

There are six sides to every dice and this is evolution's throw of confusion:

Mutations (Always destructive)

Randomness (Produces disorder)

Suppositions (The guess work factory)

Denial of facts (Abandonment of reason)

Mathematical impossibilities (Denial of reality)

Chance (No intelligent results)

Life is busy for most humans, and there is much information in the public domain for cognitive consumption on this topic. I could give many thousands of scientific facts but it possibly would be overload! What I wish to concentrate the mind on is ONTOGENY: the moment of conception to the moment of birth. This mystifying process is staggering in its teleological end results growing imperceptivity from a seed/sperm smaller than a pinhead. ***Please Stop & Consider:*** one the greatest joys humans can experience is the moments, hours, days after a baby is born. Race, colour, religion, rich or poor matters not. The mother beams with joy at her helpless little baby, and the father is usually filled with amazement at the child, with unbounded feelings of love and admiration for his woman. Sometimes no words are spoken as the mother and father gaze in wonder at their offspring and occasionally lock eyes as if looking into the very soul of fathomless, mysterious, magnificent beauty. These are moments of deep love and happiness that time cannot erode nor money buy. It is treasured glimpses of magical moments like these, that memories of bliss burn into hearts and minds. It is then reflective intelligence wonders and silently whispers expressions of thanks and asks questions of ultimate purpose.

For about 6,000 years people have pondered, questioned, analysed and commented on the birth process. However with the relatively recent invention of very powerful microscopes and cameras we can peer into the peerless world of incipient pregnancy and marvel at the scientific genius of supreme biological orderliness unfolding before our very eyes. Read Lennard Nilsson's powerful book, *A Child is Born,* and marvel at his marvellous pictures of the growing embryo and foetus. . . . to the triumph of passions helpless sweet joyous fruit.

In the space of ten weeks every single part of the human body, male or female is now fully developed and is the size of our thumb. Look at your thumb and imagine you and I were that size and ten weeks previously the size of a pinhead! (After fertilization)

Focus on this and we start to comprehend the scintillating scientific certitude of a **Cognitive Architectural Genetic Engineer** (**GOD**). There is simply no other credible explanation . . . for such complex interconnected biological growth. When the magnitude, majesty, mystery, metabolism and mellifluous mightiness of the growing embryo dawns we begin to appreciate this powerful mitosis is unstoppable, till the day when its little heart beats four times faster than its mothers, in its fight to be born . . .

Reproduction is staggering in complexity and wondrous in form, how from tiny seeds all the:
- Fish in the sea.
- Bird's in the sky.
- Animals on the ground.
- Fruits on trees.
- Vegetables in soil.
- Flowers in bloom.
- Babies in bellies.

All possess within them the mechanism to reproduce themselves ...

From this BLASTOCYST

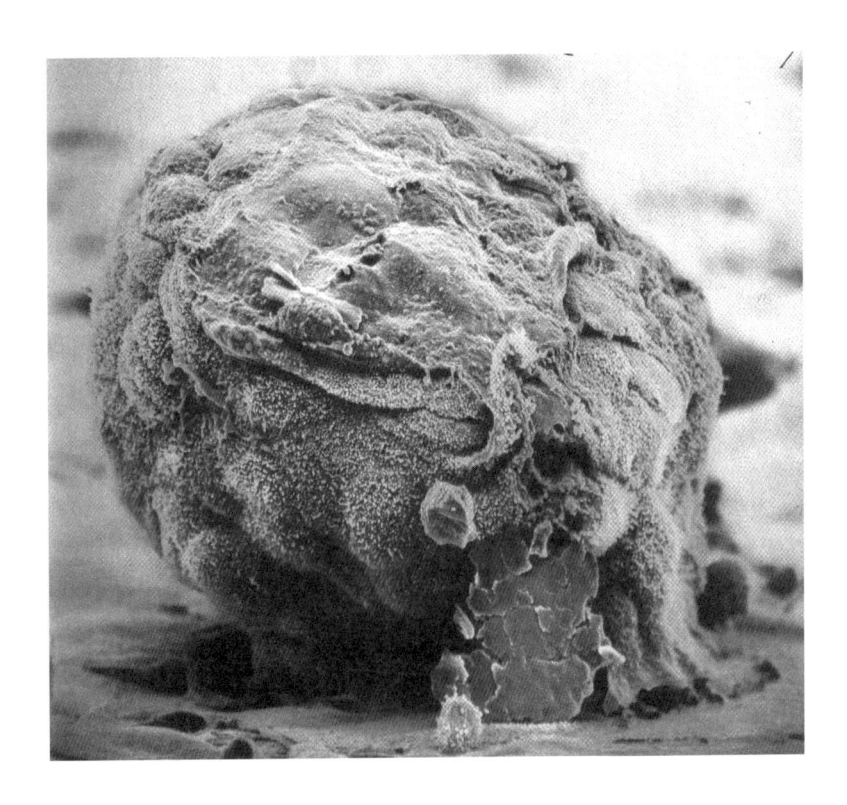

To this BABY

BABY WITH PARENTS

The staggering amount of genetic codes encapsulated in the above blastocyst mocks any scientist who claims that chance selection, chance mutations and chance origins produced a living cell, a baby and an infant's astonishing tiny reproductive system. It simply is not credible and *those who preach non-sense have lost good sense.*

BEGINNINGS

Each of us developed from a genetic blueprint with a staggering amount of inter-related, interconnected biochemical cells all in their correct place with each one being predetermined by DNA. The human *Genome* we now know has over three thousand million chemical letters. The amount of instructions in each cell is virtually incomprehensible, however for this book it is enough to say in plain English that no amount of bluffing, guessing, suppositions, conjecture, surmising or wishful thinking among confident con-artists can disguise the scientific facts.

I don't care if a person does or does not believe in God! Only please do not try and fool people with evolution's CLAP-TRAP! I don't care if anyone wishes to ignore the provable scientific facts, that's your choice, only don't try lay un-provable, fictitious, factitious dross on others! Belief in evolution is akin to 'Faith in Fairies.' (Actually there is more truth in fairies since there is a secondary dictionary definition when used as offensive slang for a homosexual man!)

Now, back to reproduction within our species. Teenagers generally do not think about replicating themselves. What they think about is friendship, smiling, flirting, attractiveness, holding hands, kissing the opposite sex. Many of us remember our first delicious kiss yet have difficulty recalling in detail our first sexual experience. Oh, yes, we remember, of course, but somehow the thrill and excitement of that first magic kiss lingers, lovely longer, in our memories!

1,000 SPERM A SECOND

A healthy young male produces 1000 sperm a second. Each sperm contains an entirely unique selection of the father-to-be's genetic material. At ejaculation 400 to 500 million sperm race madly to reach, penetrate and fertilize one egg released out of 500,000 latent eggs in the woman's body since she was a five month old foetus. One out of 500,000,000 (five hundred million) wins the prize of life when it enters the female egg which is one out of 500,000.

500,000,000   Sperm pleasurably released at each ejaculation

500,000   Eggs over a lifetime (safely protected in ovaries!)

1 Egg mysteriously released every month

Each released egg contains 23 vital chromosomes necessary to make a baby.

Each sperm contains 23 vital chromosomes necessary to make a baby.

Every single day young males produce 100,000,000 (1 hundred million) sperm.

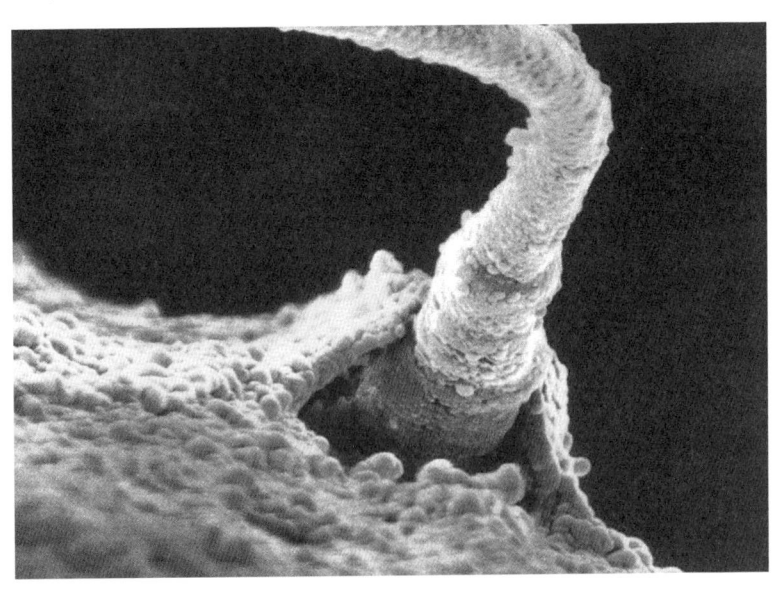

THE WINNING SPERM!

Every single one contains, yes every single wiggly sperm, hosts the father's genetic code.

A blueprint for a supersonic jet aircraft or a rocket designed to land on the moon takes years of intelligent purposeful planning, coupled with magnificent mathematics to produce the desired result. Only a fool would accept those plans being the result of blind chance. Yet many educated people caught in the vortex of delusional, specious scientific psychobabble plus the crossfire of religious fundamentalists and vocal atheists draw conclusions alien to reason and foreign to facts.

For example in one of the many books I have read purporting to demolish a **Cognitive Architectural Genetic Engineer—GOD)** possibly the worst was by a man called Michael Shermer who wasted his time writing *Why Darwin Matters* (to him!) published 2006. Not once in the 190 pages did he present one unambiguous, demonstrable scientific example of the formation of a new species by the accumulation of mutations. The reason being simple—there are none as has been so cogently challenged to many in the deluded scientific community by Lynn Margulis, a distinguished Professor of Biology at the University of Massachusetts.

She is highly respected for her widely accepted theory that mitochondria, the energy source of plant and animal cells, were once independent bacterial cells. And Lynn Margulis says that history will ultimately judge neo-Darwinism as "a minor twentieth century religious sect within the sprawling religious persuasion of Anglo Saxon biology." At one of her many public talks she asks the molecular biologists in the audience to name a single, unambiguous example of the formation of a new species by the accumulation of mutations. Her challenge still goes unmet. Proponents of the standard theory, she says, "wallow in their zoological, capitalistic, competitive, cost benefit interpretation of Darwin- having mistaken him . . . Neo Darwinism, which insists on (the slow accrual of mutations), is in a complete funk."(State of panic)

Michael Shermer's book published (2006) by Henry Holt and Company, Inc., *Why Darwin Matters* confuses the intellectual architecture of our universe, earth and everything in it with religion. He conveniently side steps the three legged stool of fantasy that evolution rests on.

Spontaneous Generation—Demolished by Pasteur

Random Mutations—Always damaging

Natural Selection—Devious hoax (Demolished by scien tific facts)

Without a shred of factual evidence Shermer blithely writes in his phantasmagorical prologue, "Of course we didn't evolve from modern apes: Apes and humans evolved from common ancestor's who lived nearly seven million years ago!"

Reader always, always remember the following . . . when a person tells you an incident that happened in someone's life seven, 17, 27, 37, 47 or more years ago you might wonder if certain details are obscure and some information muddled.

When they try to tell us exactly what happened on earth 3,000, 4,000, 5,000 or 6,000 years ago they had better present some verifiable evidence! However when they pontificate on what took place millions of years ago, take it with a pound of salt, for they truly do not know, since they were not there and are simply speculating, usually using arcane language to strengthen their weak, in fact hopeless case!

Do I know what happened seven, 77 or 777 million years ago? No, of course not and to write or say I did would be dishonest, speculative and ridiculous. Nobody knows for sure, and a little more honesty would mean a lot more people would have a little more respect for a lot more scientists.

7 REASONS WHY:

They are guessing.

Most evolutionists disagree with each other.

They don't know what they are talking about.

Nobody knows for sure.

The very ancient past is a mystery.

They ignore the scientific immutable law: Biogenesis.

Time is destructive not constructive.

His first chapter is delusionally titled "The Facts of Evolution." Yet none, yes, none are produced! He publishes a known lie in scientific circles by stating, "The change from one species to another, however, happens relatively quickly on a geological time scale and in these smaller geographical isolated population groups (punctuated)." He can write any drivel he wishes but where are the facts to prove those imaginary acts? This painful nonsense gets worse when he laughably states the following: "In fact (he shows none) species change so rapidly that few "transitional" carcasses create fossils to record the change." One has to think about this to see the utter stupidity of that proposition. The fact is that there were no transitional fossils found in Darwin's time or before, and 150 years later still not even one unambiguous example has been unearthed! Even Darwin would be embarrassed by this, but at least he was honest in his prehistoric ignorance compared to known scientific facts today.

N.B: For one species to "CHANGE" into a different species—BILLIONS of chemical letters would be necessary . . . I leave you the reader to think of the implications of that. . . .

(Clue: It is the equivalent of postulating a frog became—The Prince)

To conclude, I expose a lie, a fiction and a fabrication on page 14 where he boldly writes: "The tale of human evolution is revealed in a similar manner (although here we do have an abundance of transitional fossil riches), as it is for all ancestors in the history of life." Does he take us for fools? Reader, beware. He showed no proof, named no museum where an unambiguous fossil in an indisputable transitional form can be viewed examined and agreed on. He did not because he cannot. And so it goes with all such books.

THE SEVEN RULES OF FICTION

Invent an idea.

Write a story.

Embellish the truth.

Don't let facts get in the way.

Create a character.

Weave a fantasy.

The more unbelievable the better.

And finally when people postulate about past millions or billions of years ago YOUR GUESS IS AS GOOD AS THEIRS in fact it might even be better!

LIFE—DEATH

When a person is told by a doctor they have an incurable disease and shall shortly die, everything changes. Everything. Feelings of invincibility, security, plans, hopes, desires, hates, loves, all change irrevocably. Nothing. Nothing remains the same. Suddenly what seemed important now appears, utterly insignificant and an impenetrable blackness envelopes the hurting frightened confused mind. Now, finally the thinking processes concentrate on true issues and real questions.

The dying human is unencumbered by trivialities and when alone is seriously reflective on destiny, meaning, purpose, life, death, randomness and cause. Never again can that person hope for a meaningful future because life is cut short and running out: what takes the place of daily struggle is lonely night-time, nightmare questions that are generally ignored when living and pondered while dying.

SEVEN QUESTIONS OF ULTIMATE MEANING:

Is there a Creator?

How did the universe get here?

Why is there an earth?

Why do humans die?

Where do we go when we die?

If there is a God, what is his purpose.

Shall I ever live again?

Answers to those interesting questions satisfy my intellectual curiosity and are based on science, reality, facts, common sense, Genesis, logic and a book called Mark . . . I am neither Jew nor Christian or a member of any detestable organised disaster called religion! I am free and I am me . . . I once was a bishop / elder in a "Christian" denomination I am ashamed to say but I would have preferred to have been a Rabbi!

Life is everything and losing it is not, I repeat, not natural. In fact it is the worst catastrophe that can happen any reasonably happy person! Living is precious and it is the inception of conception that decisively demolishes the apocryphal delusion called evolution.

ONTOGENY

This is the "Miracle of Miracles" where about 500,000,000 sperm on spring-load release race to find the waiting egg. Nothing is left to chance. For 6,000 years men and women have mated and had absolutely no idea how pregnancy took place. The act happened, periods were missed, female stomach's expanded and after eight to nine months a baby was born!

One fertilised egg/cell, twelve to twenty four hours later miraculously splits (see cover of book) and suddenly there are two cells. Inside this astonishing chemical factory a thousand times smaller than this " . " is a future president of the United States, possibly an astronaut, a brilliant runner or whatever . . .

The point being in layman's terms: inexorably—scientifically—mathematically—genetically linked hundreds of millions of growing cells, with each and every single one knowing where it is to go, where it is to remain, what it is designed to become, how much to grow and when to stop growing, what colour of eyes, hair, skin, how tall, small, fat or slim: Each baby has a chemical time clock with ten tiny specks in top and bottom gums which about six months later start to grow into 20 perfect symmetrical white teeth. (Imagine if newborns were born with teeth how very sore nipples would be!) About six years later the clock starts again—milk teeth fall out and thirty two new teeth for life grow from under gums. The chemical clock strikes again about seven years later when girls start to develop breasts and are slowly turning into women and boys' voices break, as they approach manhood, both preparing for adulthood incorporating possible parenthood.

All this cellular and chemical development could not happen if the following seven dangerous elements took place:

Chance.

Mutations.

Blind selection.

Chemical accidents.

Hazard.

Randomness.

Purposelessness.

Every single square inch/millimetre of our bodies from the hair on our heads to the nails on our toes and everything in between screams architectural biological design. I don't care what anyone believes but at least bring common sense to the discussion. For example paediatrics is the branch of medicine that deals with the care and treatment of infants and children. Specialists might chose the discipline of hearts, ears or eyes and spend their lives studying that particular field and yet never ever know it all.

From this dot " . " grows a complete human being with all organs functioning perfectly, each and every single one precisely interconnected, interrelated, intermarried, and generally nothing chemically interferes with the awesome final result . . . A living breathing beautiful bouncing helpless lovely baby!

Each organ and system depends on all the others to function correctly and contribute to the physiological, psychological and emotional balance that makes us human.

I am not qualified to write a medical textbook! However below are seven photos that are mesmerising in construction, complexity and thorough completeness. All are vital to healthy living and the collapse of one will unalterably cause major problems.

THE ORIGIN OF SPECIOUS NONSENSE

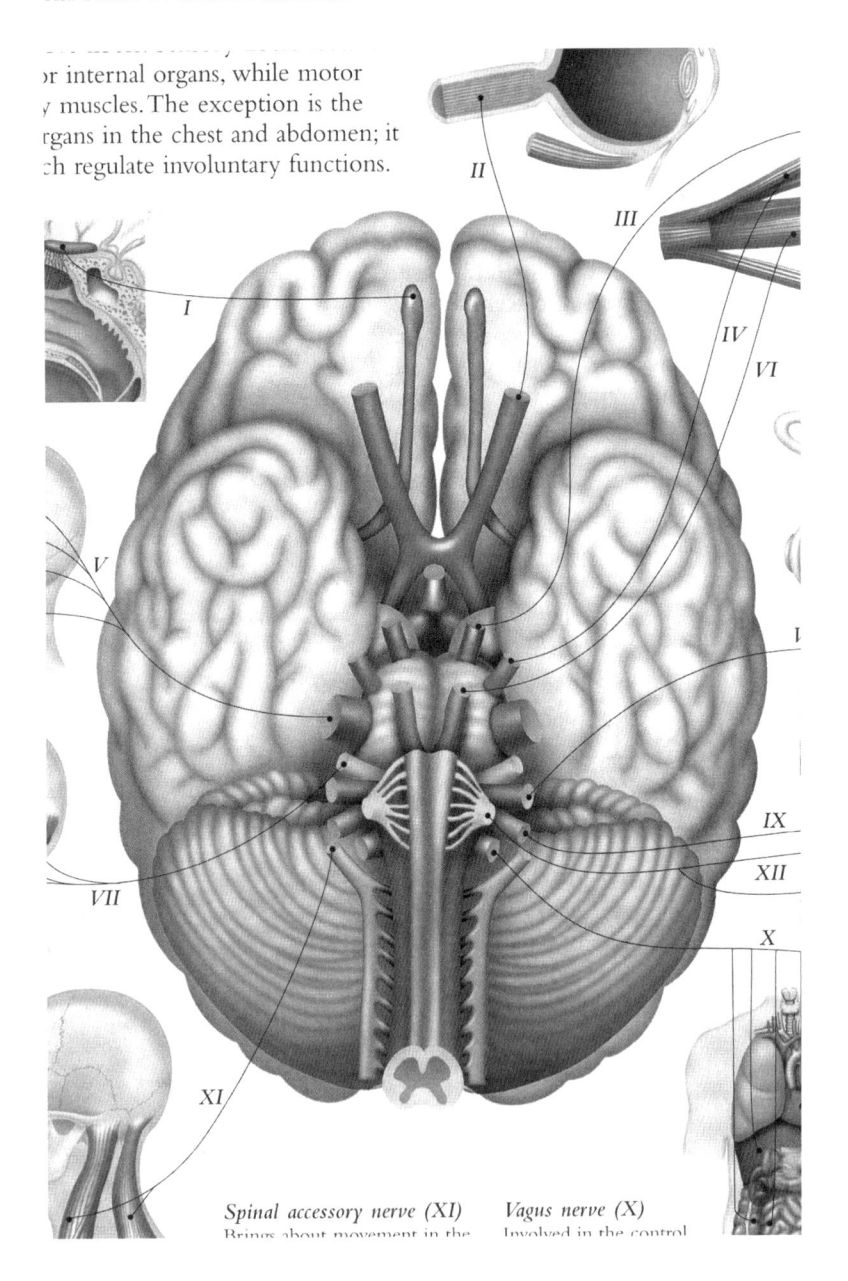

…or internal organs, while motor
… muscles. The exception is the
…rgans in the chest and abdomen; it
…h regulate involuntary functions.

Spinal accessory nerve (XI)
Brings about movement in the…

Vagus nerve (X)
Involved in the control…

BRAIN

JOHN J MAY

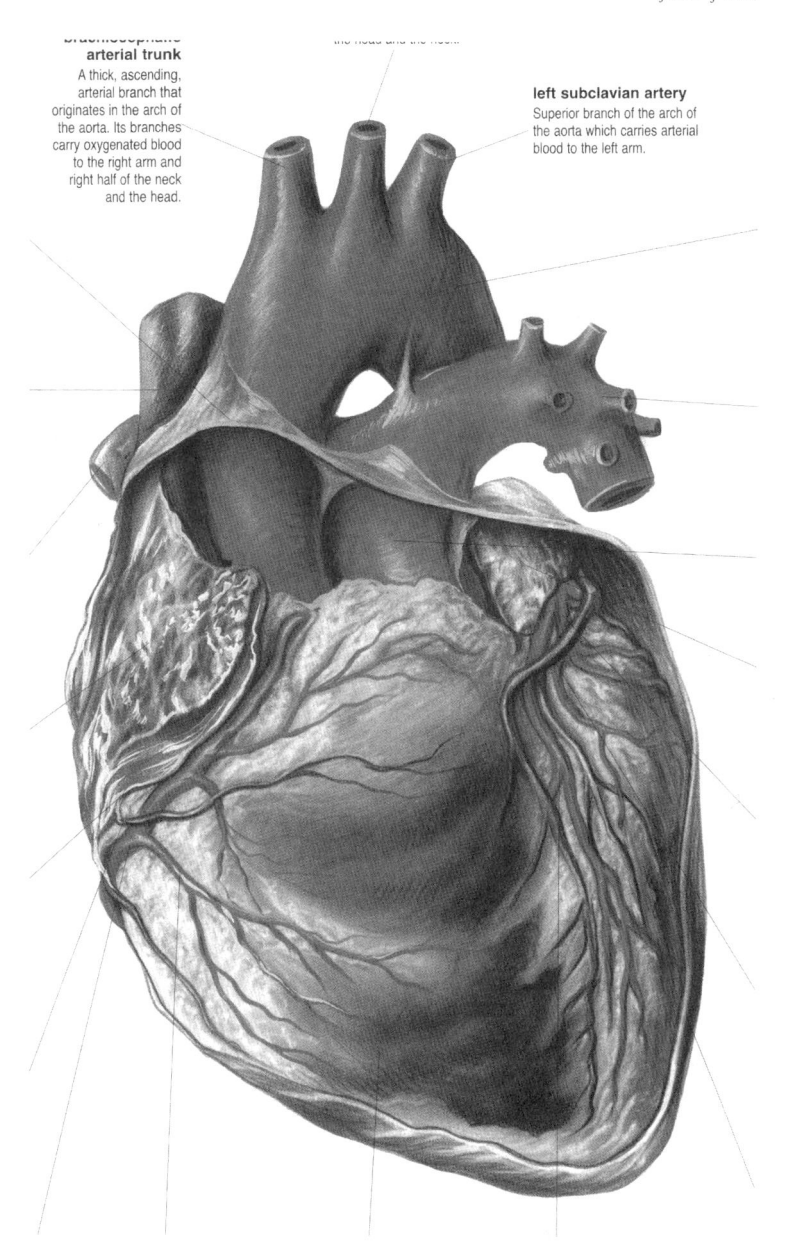

brachiocephalic arterial trunk
A thick, ascending, arterial branch that originates in the arch of the aorta. Its branches carry oxygenated blood to the right arm and right half of the neck and the head.

left subclavian artery
Superior branch of the arch of the aorta which carries arterial blood to the left arm.

HEART

THE ORIGIN OF SPECIOUS NONSENSE

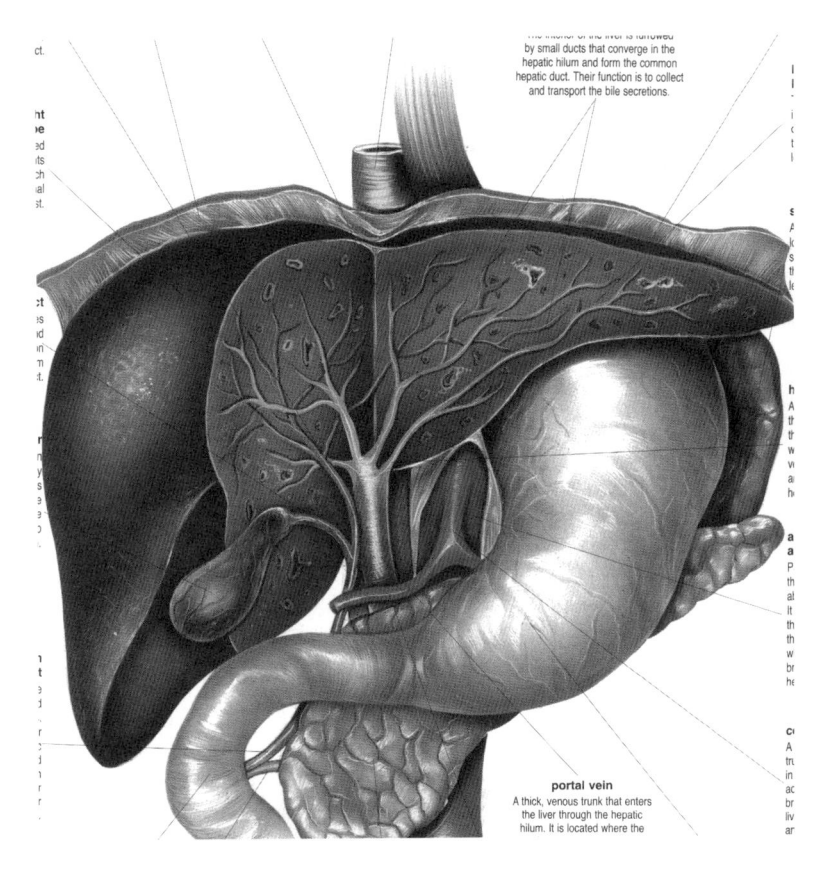

The interior of the liver is furrowed by small ducts that converge in the hepatic hilum and form the common hepatic duct. Their function is to collect and transport the bile secretions.

portal vein
A thick, venous trunk that enters the liver through the hepatic hilum. It is located where the

LIVER

JOHN J MAY

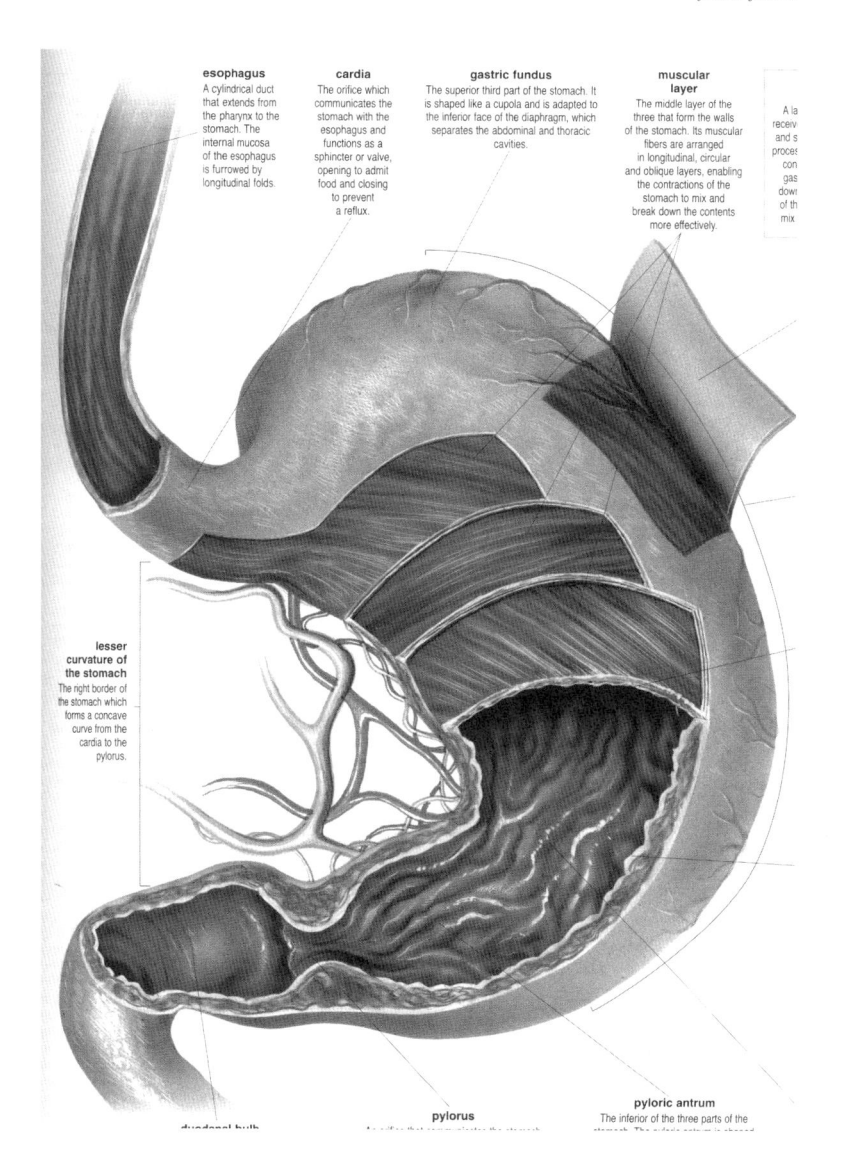

esophagus
A cylindrical duct that extends from the pharynx to the stomach. The internal mucosa of the esophagus is furrowed by longitudinal folds.

cardia
The orifice which communicates the stomach with the esophagus and functions as a sphincter or valve, opening to admit food and closing to prevent a reflux.

gastric fundus
The superior third part of the stomach. It is shaped like a cupola and is adapted to the inferior face of the diaphragm, which separates the abdominal and thoracic cavities.

muscular layer
The middle layer of the three that form the walls of the stomach. Its muscular fibers are arranged in longitudinal, circular and oblique layers, enabling the contractions of the stomach to mix and break down the contents more effectively.

lesser curvature of the stomach
The right border of the stomach which forms a concave curve from the cardia to the pylorus.

pylorus

pyloric antrum
The inferior of the three parts of the

duodenal bulb

STOMACH

125

THE ORIGIN OF SPECIOUS NONSENSE

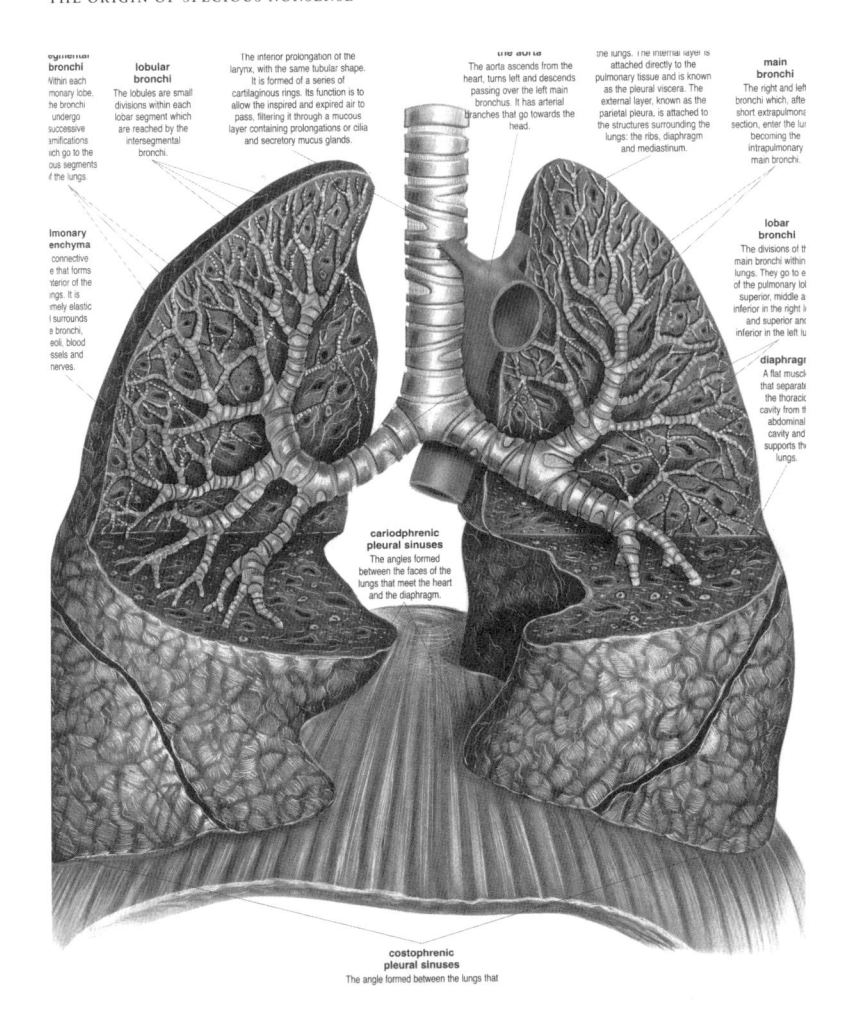

egmental
bronchi
Within each
monary lobe,
he bronchi
undergo
successive
amifications
ich go to the
ous segments
f the lungs.

lobular bronchi
The lobules are small divisions within each lobar segment which are reached by the intersegmental bronchi.

The interior prolongation of the larynx, with the same tubular shape. It is formed of a series of cartilaginous rings. Its function is to allow the inspired and expired air to pass, filtering it through a mucous layer containing prolongations or cilia and secretory mucus glands.

the aorta
The aorta ascends from the heart, turns left and descends passing over the left main bronchus. It has arterial branches that go towards the head.

the lungs. The internal layer is attached directly to the pulmonary tissue and is known as the pleural viscera. The external layer, known as the parietal pleura, is attached to the structures surrounding the lungs: the ribs, diaphragm and mediastinum.

main bronchi
The right and left bronchi which, afte short extrapulmona section, enter the lu becoming the intrapulmonary main bronchi.

lmonary
enchyma
connective
e that forms
erior of the
ngs. It is
mely elastic
l surrounds
e bronchi,
eoli, blood
ssels and
nerves.

lobar bronchi
The divisions of th main bronchi within lungs. They go to e of the pulmonary lol superior, middle a inferior in the right l and superior and inferior in the left lu

diaphragi
A flat muscl
that separate
the thoracic
cavity from t
abdominal
cavity and
supports th
lungs.

**cariodphrenic
pleural sinuses**
The angles formed between the faces of the lungs that meet the heart and the diaphragm.

**costophrenic
pleural sinuses**
The angle formed between the lungs that

LUNGS

126

JOHN J MAY

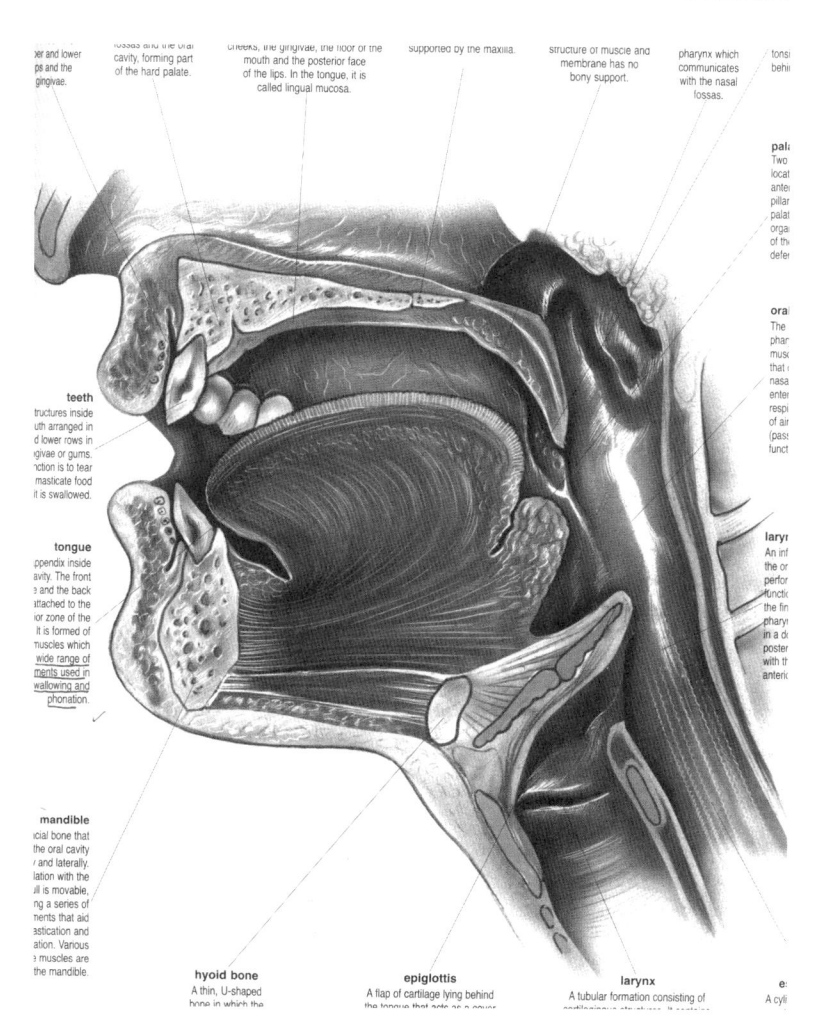

er and lower ossas and the oral cheeks, the gingivae, the floor of the supported by the maxilla. structure of muscle and pharynx which tonsi
ps and the cavity, forming part mouth and the posterior face membrane has no communicates behi
gingivae. of the hard palate. of the lips. In the tongue, it is bony support. with the nasal
called lingual mucosa. fossas.

pal
Two
locat
anter
pillar
palat
orga
of th
defer

oral
The
phar
musc
that
nasa
enter
respi
of ai
(pass
func

teeth
tructures inside
uth arranged in
d lower rows in
givae or gums.
ction is to tear
masticate food
it is swallowed.

lary
An inf
the or
perfor
uncti
the fir
phary
in a d
poster
with t
anteri

tongue
ppendix inside
avity. The front
e and the back
ttached to the
or zone of the
It is formed of
nuscles which
wide range of
ments used in
wallowing and
phonation.

mandible
cial bone that
the oral cavity
y and laterally.
lation with the
ll is movable,
ng a series of
nents that aid
astication and
ation. Various
muscles are
the mandible.

hyoid bone
A thin, U-shaped
bone in which the

epiglottis
A flap of cartilage lying behind
the tongue that acts as a cover

larynx
A tubular formation consisting of

e
A cyli

THROAT

127

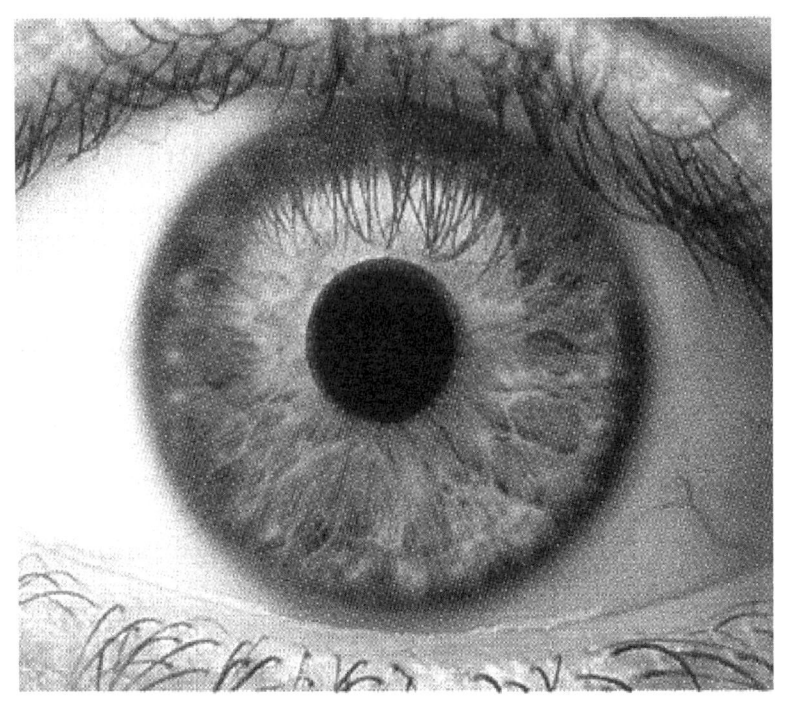

EYE

If a man or woman lived a million years there would not be enough time to examine and understand completely one single cell and our bodies possess one hundred trillion with each and every one a testament to chemical manufacturing, and economic reproductive brilliance.

There are thousands of books written about the human body, its complexity functions, parts, and what it takes to keep it safe in a clean and healthy manner. Most authors agree on its genetic, staggering intricacy and in particular the molecular beauty of our eyes and the astonishing inter-related way they function. I write and you read because of vision.

Why am I even writing this? Why do I care if people believe in God or not? Well I shall tell you. There are seven things I truly detest with a shaking passion:

War.

Evolution.

Organised religion.

Lies.

Haughtiness.

Violence towards women.

Hurting children.

War is caused by madmen who care nothing for the feelings and rights of others. It is rationalised by nationalism and sanctified by various rotten religions.

Evolution is a toxic poison that destroys innocence, abandons reason, ignores science, corrupts children, and turns the clinically violent into megalomaniacal destructive cruel monsters who believe they are answerable to no one. The psychotic mindset in those types of people is legitimised by the writings of many societal innocents in the scientific community and sanctified by all the major sects. In their sociopathic delusions, their particular tribe are in a struggle for survival and the fittest shall triumph regardless of how vicious, cruel and flagitious their tribal actions become. After all, in a convinced evolutionists mind we are only a higher form of animal, the result of millions of years of violent struggle. And, yes this poisonous, dreary theory sickens me. . . .

Organised religion is a disaster throughout history. It hardens hearts against our fellow-man, burdens lives with fictions, torments minds with unbelievable fantasies and breeds supercilious contempt for other peoples sincere beliefs. All civilised educated societies should stop supporting these tax free shelters.

Lies. It hurts to be lied to! Ask a child or a woman who has been deceived by the man she loved (or visa versa). It is in every sense putrid and when discovered, trust like a pane of beautiful stained glass smashed to smithereens. How again can a practiced liar ever be trusted? It's possible but exceedingly rare and only happens with a change of heart.

Haughtiness: The American Declaration of Independence on its first page signed by 56 noblemen from thirteen states dated July 4, 1776 states: "We hold these truths to be self evident, that all men are created equal." Is there anything more revolting than one human being looking down on another because of race, religion or social standing? It is a mirror to the true shallowness of the false soul.

Violence towards Women: Whenever I hear of women being beaten (practically a pastime in some Muslim countries), my skin crawls with an inward rage and I wish the cowards who do it, could in some way feel the awful pain they inflict. I have four daughters and if ever one of them was injured by a coward I know I would seek justice with a passionate vengeance. Yesterday in my city, Dublin, Ireland; a 46 year old woman was beaten to death. So cruel. So sad. So savage. So tragic. So cowardly . . .

Hurting Children: Every generation has produced hurt bewildered and marginalised damaged children. The causes are many but the results are the same: A violent childhood tends to replicate itself in adulthood.

The United Nations; Universal General Assembly Resolution 217A of December 10th, 1948 states:

Article 1:
"All human beings are born free and equal in dignity and rights.
　　　　They are endowed with reason and conscience
　　　　and should act towards one another in a spirit of
　　　　brotherhood."
Article 25:
(1) "Everyone has the right to a standard of living adequate for the
　　　　health and well-being of himself and of his family,
　　　　including food, clothing, housing, medical care and
　　　　necessary social services, and the right to security
　　　　in the event of unemployment, sickness, disability,
　　　　widowhood, old age or other lack of livelihood in
　　　　circumstances beyond his control."
(2) "Motherhood and childhood are entitled to special care and
　　　　assistance. All children, whether born in or out of
　　　　wedlock, shall enjoy the same social protection."

The beating of children is child abuse and unbelievably encouraged by some holy books, with sticks! Sad but true, and yet a famous Jewish rabbi scolded his ignorant, fussy followers 2,000 years ago and warned anyone who hurt children they deserved to be drowned. Children worship their parents and guardians and it is always wrong to bully, hit, batter or bruise them.

If there is a God/Creator/First Cause/ could he not have, somehow, been more solicitous in revealing his unalterable unambiguous unimaginable universal unfeigned unfailing wonderful purpose? Has he in some way been derelict in informing mankind of his kind purpose? It would appear so! My book shall hardly dent the ignorance we were all born into. If the Bible is supposed to reveal his purpose (which it certainly seems to) then why are 99% of the world's population ignorant of its contents? In fact it's only until very recently in history that ordinary workers had access to education and books, producing mass literacy as opposed to mass ignorance on a truly massive scale.

As we progress technologically at speeds approaching light, the increase in knowledge for manufacturing weapons of hugely destructive capacity, will surely come into the hands of terrorists who think destruction, killing and destroying, on a huge scale is in some way pleasing to their God, Allah, or whoever! Holy Books, Holy War, Holy Gods and Holy Terror shall surely combine to produce an apocalyptic holocaust! What is the seed of such horrifying holy hatred? It's partly historical with the mix of holy books and unholy leaders seeking holy revenge!

JOHN J MAY

MANS FLAGITIOUS CRUELTY

CHOICE REGARDING EVOLUTION

I write for those who do not know what to believe! Wise decisions are the bedrock of quality in our lives. Depending on what we decide indubitably affects our sleep, peace, restful hearts and love lives.

Seven Decisions

To accept friends faults!

To be faithful or unfaithful!

To gossip maliciously or refrain!

To yield to temptations or refuse!

To indulge passions or reject!

To believe evolution or reason!

To acknowledge the strong possibility of a creator!

Once evolution is accepted in the mind the heart might cease being kind! Why?

Because if we are only a higher form of animal, what does it matter? Who cares? It's our attitude to life that determines life's attitude toward us. Things that matter most must never be at the mercy of things that matter least! If there is a God there may be a better life for the human family one day and that includes the dead! Ludwig Wittgenstein (1889—1951), an Austrian philosopher, whom some view as the 20th century's most important one, spoke of the "limits of language and meaning." He had this to say on the idea of resurrection.

"If a God exists then a future resurrection is quite likely!"

If we exist in a cold black galaxy and our milky way is designed in an orderly fashion and earth teems with beautiful wonderful species surely it is logical to deduce there is a vastly superior intelligence behind everything! If he decides one day to take over control of his planet from man's inept attempts at governing then so be it! It is his property his purpose and his prerogative. It would be as easy for him on a molecular level to resurrect back to this earth Abraham, Isaac, Jacob and their three wives if he so wanted. In fact it would be easier for him to do that than for you and I to make a cup of coffee!

Our world is filled with fear, our future uncertain and made worse by the toxic bleakness and weakness of an idea that does not even deserve to be called a theory -evolution! The word THEORY means: "A PROPOSED EXPLANATION WHOSE STATUS IS STILL CONJECTURAL, IN CONTRAST TO WELL ESTABLISHED PROPOSITIONS THAT ARE REGARDED AS REPORTING MATTERS OF ACTUAL FACT." For example it is an established fact that the earth is round. It is also an established fact that scientists cannot agree on how life and the universe originated. Some 99.9% of all scientists agree the earth is round and 99.9% disagree among themselves as to the precise time and mechanisms of life's beginnings. And the reason for the speculative rantings, musings and guesswork is ever so simple: THEY DO NOT KNOW! When they write their speculative imaginations in National Geographic and Scientific American (both excellent magazines) always remember: they do not know what they are talking about!!! And that is a fact. I could quote a thousand scientists worldwide who reject evolution for the deceit and hoax it is, but two are sufficient:

1. Ms Lynn Margulis is a distinguished University Professor of Biology at the University of Massachusetts: At one of her many public talks she asks the molecular biologists in the audience to "name a single unambiguous example of the formation of a new species by the accumulation of mutations." Her challenge goes unmet. Proponents of the standard theory she says, "wallow in their zoological, capitalistic, competitive cost benefit, interpretation of Darwin—having mistaken him . . . Neo-Darwinism which insists on the slow accrual of mutations, is in a complete funk." (Funk means "state of panic.")

2. The second quote I have chosen (among thousands from anti-evolutionist scientists) is from: Dr. T.N. Tamisian of the United States Atomic Energy Commission. "Scientists who go about teaching that evolution is a fact of life are great con-men, and the story they are telling may be the greatest hoax ever. In explaining evolution we do not have one iota of fact." Taken from *Evolution and The Emperor's New Clothes,* Roydon publications 1982. The American Heritage College Dictionary (third edition) under FACT states; "something having real, demonstrable existence." Also . . ."The quality of being real or actual." That being a fact and me being a factualist, then I shall spend my life exposing scientists who advocate fantasy as fact!

Evolution cannot be unambiguously demonstrated in any laboratory worldwide. It could not in Darwin's day, today or in the future, because on a molecular, biochemical, cellular DNA and atomic level, it is an utter scientific impossibility.

The chemical structure of cells in all, I repeat, all species is staggeringly complex, strategically imprisoned and enough to say no scientist in the world understands the inner amazing interdependent molecular maze of even one cell and humans have trillions of them.

Around 200 different types of cells have been identified in the human body. The average man has approximately 25 trillion red blood cells which carry vital oxygen around the body. This is an estimated one third of the total number of cells in the human body of around 75 trillion—each cell is a busy factory (or city) carrying out several thousand different tasks in an orderly, integrated way, and yet each and every one has a specialised role . . .

We all started as one tiny living cell (see cover of book) and were chemically constructed by the same anatomical blueprint with a few nice fixed differences between the sexes. There is no machine, rocket, airplane, computer or anything you or I can think of to equal the mathematical—biochemical genius, of the male and female body which in a very beautiful way—love—reproduces itself starting once again from one tiny sperm out of 500 million and one tiny tremendous egg out of 500,000 . . .

Shame on educated men and women who see all this interconnected beauty, complexity, obvious architectural design, mesmerising assembly of purposeful components which repairs and replicates itself and say, "See this—this all came about by mutations mistakes and careless chance." Oh yes, shame on you!

Evolution is corrosive to science, blinds people to reality, teaches acceptance and social acquiescence, numbs reason, stultifies intellectual discussion and is fatuously taught by some, adopting an air of supercilious smug confident certitude. In their arcane delusional language they have sacrificed 'Reason' on the altar of 'Ideas' and so few have "The courage that dare not speak its name." Their speculative specious spurious shallow sham shall one day be exposed for the entire world to see. It will be done when courageous scientists say "Enough of this unscientific hogwash! Let's promote the facts!"

I detest organised religion much more than the incisive intelligent educated Richard Dawkins, Daniel C. Dennet, Sam Harris and many others whose intellect I admire and respect and whose books I have thoroughly enjoyed reading. But enough of the pretension that evolution is an established fact—when nothing could be further from the truth! Because some religious leaders make fools of themselves in this ongoing debate, do not make the mistake of coupling me/us with them, for I tell you; that in particular would be the rock on which you perish!!! (Metaphorically writing of course!!)

Evolution is detestable as it is a form of child abuse. It is an idea whose time has past! It is degrading to our intelligence, dishonouring to our species and in the final analysis is an insulting fiction towards our maker.

SCIENCE V THEORY

The definition of Science is: "Learning or study concerned with demonstrable truths or observable phenomena, and characterised by the systematic application of scientific method."

The definition of Theory is:
"A system of assumptions,"
"Abstract reasoning; speculation,"
"An assumption based on limited information or knowledge; a conjecture."

A canard is a deliberately misleading story like the fairy tale of evolutions miasma. It has stultified and poisoned minds of otherwise intelligent people and reduced them to followers much like religions reckless renegades against reason.

For example, a molecule is a group of atoms held together by chemical forces. In the atomic structure of a newly fertilised human female egg, concrescence takes place; this is the chemical orderly growing together of millions-billions-trillions of related parts, tissues, cells, known as **ONTOGENY** which originates from the staggeringly complex world of ovulation plus sperm production. And it is when these two worlds of malleable, mysterious,s mammalian massive mitosis collide and cellular growth commences that life begins to organise and grow in the welcoming womb. Some 90,000 miles of veins are growing that will one day facilitate 75,000 pints of blood flowing through every strategic metre every 24 hours. Some 300 bones are growing, as well as 600 muscles and 100 billion brain cells are forming. There are 80 million red blood cells inside every cubic inch of the body and each one possesses 270,000,000 haemoglobin (iron) molecules.

The heart will beat about 100,000 times, we take 20,000 breathes, blink once every five seconds or on average about 17,000 times, *every single day* . . .

The above is nothing, compared to in depth analysis of the interconnected beautifully co-ordinated functioning growing embryo. When my daughter Judith, gave me a gift of Lennart

Nilsson's magnificent book, *A Child is Born*, I was only mildly interested! By page 75, I was hooked! Finishing page 150, I was mesmerised. On 14 April, 2008 when I finished page 204, I was an angry man!

Here at last is the scientific mechanism by which reason unites with observable reality (under microscopic observation) and injects the appreciative heart with gratitude. Here at last is incontrovertible proof of immutable scientific laws. Here at last is the dawning realisation that you and I once were smaller than the dots on these i i i's. Here at last is logical proof of a creator, God, architect, biologist, mathematician, genius. The Cognitive Artistic Genetic Engineer . . .

This war is not really between science and religion or as Chicago geneticist Jerry Coyne declared, "Between rationalism and superstition!" It is between human bias and intellectual persuasion—bias is the blood of basic convictions, the transfusion of transcendent tranquilizers. Bias breeds beastly bondage and produces millions of mental slaves, writing singing and shouting in praise of their slavery. A black cat is still a black cat as Ludwig Wittgenstein said. However, to the person who wants to debate, 'What is a cat?' nothing is real and endless philosophical debate is the interminable result.

On the other hand when we sit alone and think on life's meaningful questions, we are generally influenced by empirical evidence and logical persuasion. We are born inquisitive, naked, ignorant, agnostic and millions die the same way . . .

It is extremely difficult to give a definition of religion but most might agree it gives assent to belief in a higher cosmic power. Those who historically agree, have given some sort of servility to this Deity in thousands of desperately different and often-times, sad and shockingly heartfelt ways—and remember each act of worship is influenced incontrovertibly by childhood influences. Most people die in the religion of their birth which is a powerful testament not to the truth of convictions, but to a life long infection of bias. The first murder that took place on this planet as recorded in a 3500 year old document called Genesis was over ' the method of worship'—a conceptual conviction that can destroy minds, hearts, lives and communities in the most fearsome flagrant flagitious manner a sincere human being can cruelly imagine! Dangerous destructive childhood conceptual convictions breed bias and are the seed that germinates and chokes reason.

There is no 'true way' to love but it is always wrong to hate. The concept of God facilitates love. Evolution facilitates hate. Many, many civilised atheists will be apoplectic with anger when reading the above and rightly so for the simple reason they in their hearts have no ill-will towards their fellow-man! That is not the point! The historic facts (as opposed to innocent wishful thinking) are, there are tens of thousands of depraved, violent men on this planet who simply need to hear this fiction once; "We are only the result of blind pitiless chance, a higher form of animal, we have arrived and survived through merciless struggle and savagery, nature is cruel and so can we; there is no God, no purpose, no point, be strong and fight, dominate the weak, eliminate them and feel no pity as it is a sign of weakness!" (as the Nazis did who believed lock stock and barrel the above recipe for moral and social chaos!)

On hearing this fruit of evolutions madness, a natural selection takes place in the delirious minds of some deluded men. Here is the licence to kill, hate and destroy. The philosophical, specious, mendacious mandate for murderous mayhem. The great tragedy in all this is, the savages who wage war on their earthly brothers have many of the scientific community collaborating, colluding and collectively urging them on through their assent to this asinine hoax; evolution . . .

Arguing to the contrary is fruitless, pointless, hopeless and fundamentally dishonest. Twelve angry atheists on a jury can never be persuaded and convinced by scientific evidence, empiric deduction, facts and common sense for the simple reason; they wish to choose to believe there never was nor ever shall be a creator. (Bias!) It is a choice they are entitled to make. A choice that defies reason, denies logic, defends pseudoscience, demands acquiescence, despises questioning, denounces spirituality, deifies theories, demonizes fellow scientists who disagree and denigrates sincere Socratic students who sense the sophism inherent at the heart of this morally toxic myth!

Many closed minds have many open sewers and some open minds have some closed hearts! For example at the very core of Darwin's theory (or guesswork) is the notion; 'variation.' This is the key to understanding his imagined selection by blind chance somehow through nature! His entire fable hangs on this invisible thread. His three legged stool of Mutations, Randomness and Chance, is credited with producing the millions of species each perfectly suited to its environment through adaption to produce variation!

Most Darwinists are unaware that he once believed and actually wrote in the first edition of *Origin of Species* that north American black bears swimming through water with their mouths open collecting insects *and concluded unbelievably;* "I can see no difficulty in a race of bears being rendered by natural selection, more and more aquatic in their structure and habits, with larger and larger mouths till a creature was produced as monstrous as a whale." Such laughable drivel is beyond belief! Yet this and many, many other phantasmagorical phantoms were seriously postulated by Charles as "Evidence" of his ludicrous laughable concocted concepts. I repeat in case you might have missed his point: He wrote that bears transformed into whales! Incidentally he was so embarrassed by this palpably puerile fantasy that he wisely removed it from all further editions.

Scientific research demands at least 95% proof for a proposition to be correct. For natural selection to be seriously postulated by some scientists with 0% proof is one of this world's true mysteries and reminds me of Dr. T. N. Tammisian's quote; "Scientists—who go about teaching that evolution is a fact of life are great con-men, and the story they are telling may be the greatest hoax ever. In explaining evolution, we do not have one iota of fact." Dr. Tammisian was part of the United States Atomic Energy Commission in 1959.

Darwin was ignorant of DNA, and consequently could not have known every single species possesses its unique information plan/code in its molecular-atomic-DNA construction. This code, and nothing else, determines the shape-size-look-skeletal-instinctual-colour-etc, of that particular species. Whether it is plants-fruits-vegetables-fish-bees-birds-animals or mankind, there are inviolable scientific *laws* of immutability which are transparent around our globe. As Charles himself wrote in his chapter; 'Difficulties on Theory,' "How can we account for species when crossed, being sterile or producing sterile offspring, whereas, when varieties are crossed, their fertility is unimpaired?" For example humans are tall, small, black, white, Asian, fat, skinny with many shades in between and all within this astonishing variety can interbreed and individual offspring are the invariable result. However, if sex between humans and animals occurs, never ever is a half human and half animal the end result. Why? because species CANNOT be crossed and so it is in the animal kingdom. And that, dear readers, *that is a fact*.

The truth about origins affects everything. However there is only one certain way to know this truth and that is to have been there at the beginning and since no one living or dead was present at that moment in time (if time was even present!) then statements regarding origins are by definition spurious, specious, speculative and merely guessing, nothing more and nothing less. The next best way of knowing would be to get a visit from the maker of the universe and while enjoying a beer together for him to actually reveal how he constructed, matter, time, numbers, motion and to throw in "Why?" would be nice! Since the likelihood of that is less than zero! (except for the lunatics in mental asylums in Israel and elsewhere who know they already have enjoyed this experience on a number of auspicious occasions!) There might be a third way of knowing! A sort of divine revelation committed to writing for a lucky few!

If there is no third way, then what I am about to write is simply speculation. I have read most of the so called "holy books" and when it comes to origins their input is off-putting!

Laughable speculation from desperately sincere fantasists passes as authoritative pronouncements on origins. The worst is

Origin of Species and the best are ancient Hebrew writings and in particular, Genesis. I do not mean to support desert tribal savagery but this fascinating record of mans meaningful origins is philosophically revelatory.

Science stands for seven things;

Facts

Evidence

Proof

Demonstrable

Manifest

Observable

Truths

Evolution proposes seven contestable ambiguities as evidence.

Chance

Mutations

Randomness

Theories

Speculation

E'ons of time

Transmogrification

The choice is between;

- PROVEABLE SCIENTIFIC EVIDENCE
- UNPROVABLE PSEUDOSCIENTIFIC THEORIES

It is basically as simple as that. When lack of proof for an accepted and respected speculative, extremely popular idea is discovered it takes great courage in the scientific community to 'Break the spell'—'End the faith'- 'Expose the evolution delusion' and prove 'Myths are not great,' to finally let ordinary people realise, when those boiling bad ideas percolate into the public domain they poison everything—everything.

Once again I state, I could give 1,000 quotes from eminently respected scientists who recognise evolution as the greatest deceit in the entire history of science. Yet even then some would still choose to believe, as it has become a cause mixed with blind faith. Be that as it may I quote here only one man, Professor Louis Bounoure, former President of the Biological Society and director of the Strassbourg Zoological Museum. He contends, as reported in the advocate on March 8th, 1984, page 17; *"Evolution is a fairy tale for grown ups. This theory has helped nothing in the progress of science, it is useless."*

Santa Claus is a fairy tale for children and evolution is one for grown ups. Its no wonder the famous journalist and philosopher of the 20th century, Malcom Muggeridge anticipated the future rejection of this fiction and calls it; "One of the greatest jokes in human history."

Sensational court cases are fought, lost or won on 'facts that are persuasive' or 'beyond a reasonable doubt.' In modern democracies to undertake a prosecution without sufficient evidence would be considered the choice of a fool. In the courtroom of life a battle rages over origins. Millions claim life, earth, universe, and everything is an accidental result of time and chance, other millions claim origins were created by a purposeful designer. There is no realistic third way. It is a choice people in both camps have made and millions outside those camps are too busy with life's struggles to give much thought to it.

Depending on childhood influences we shall be indubitably attracted to one or the other! (or none!) Background bias is the teacher and choice is on automatic pilot to facilitate easy options. Thinking is difficult, decisions awkward, consequences scary, the past speculative and our future dangerously unknown. However basic truths are extremely liberating and intellectually fascinating. Michael J. Behe wrote in his profoundly interesting book, *Darwin's Black Box*, published by Simon & Schuster, Inc. (2006) "The minimal number of components necessary for a complicated machine to function is called, irreducible complexity." For example; A: Remove a cells membrane and it collapses. B: Remove mitochondria, there is no energy (Rendering it virtually useless) C: Remove the cells nucleus, most chemical functions cease. Bacterium (Mycoplasma Genitalium) is the smallest amount of genetic material known of any organism and yet has 580,000 base pairs on its 482 genes! Magnify one human cell 1000 million times until it is 20 kilometres in diameter, (the size of a modern city) order and biological brilliance of un-parallelled complexity and adaptive design becomes apparent. Millions of openings and closings are evident and supreme molecular technology of bewildering complexity is mesmerizingly manifest.

Mycoplasma genitalia has 580,000 base pairs but humans in our genes/DNA have 3 billion base pairs. All this information would fill over 42,700 large books of very small print and these books used as the template for the construction of a human being. All the information for the bodies functional systems are mathematically, molecularly and biologically interconnected through purposeful intelligent architectural design. Oh, and we have 100 trillion cells! The point of the above is to make a scientific factual statement: DNA cannot form by chance. If an encyclopaedia cannot be the result of chance, then, as Sir Fred Hoyle the Eminent British Astronomer cogently pointed out;

"It is 1 to the number 40,000 zeros that was calculated for life to have arisen from non living matter!" In other words it is an utter impossibility.

A simple illustration of chance is to get 15 tennis balls and throw them up in the air and expect them to form a perfect 360 degree circle! It's never going to happen!

Or get 1 million blind people and give them 1 million Rubiks cubes and tell them they must all start at the same time and finish precisely in seven hours all at the same time! *It's never going to happen.*

Or fill a standard gym with 100,000 dice and set off an explosion and expect every single one of them to land on a six! *It is never going to happen!*

Or go to a massive airplane graveyard sited in the path of tornados with the expectation that when one hits it will amalgamate the junk somehow and produce a Boeing 747! *It is never going to happen.*

Or the odds for spontaneous generation is mathematically 1 to 340,000,000 zeros! *It is never going to happen.*

The eminent Dr. Carl Sagan estimated the mathematical probability of the (so called) simplest forms of life emerging from non-living matter to be 1 x followed by 2 billion zeros! *It is never going to happen.*

Dr. Emile Borel who is credited with discovering the laws of probability says *"1 x 50 zero's never happens."*

Commenting on the myth of the "Prebiotic Soup," the noted molecular biologist, Michael Denton said, "It comes as a shock to realise there is no—absolutely no evidence for its existence." Therefore, the logical corollary of this fact is: *promoting it is scientifically indefensible, inexcusable, and insanely inane.*

To build the story of life on a theory of chance—randomness and mutations is like building a house on sand. Some 99.9% of all mutations are harmful and they must occur in the genes of reproductive cells, to be genetically effective. However genes are generally very stable and those cells have enzymes which constantly monitor on a molecular level, checks and safeguards. The world, life or universe of the cell is so utterly complex with millions of functions taking place every heart-beat and all beautifully interconnected—all somehow knowing what to do; when to do it; where to go in the human, animal or vegetable structures; what chemicals to produce—exactly where to place them; how to safely remove wastes and toxins; how to identify them—plus tens of thousands of beneficial programmed decisions every second to keep the organisms clean, healthy, attractive and reproductive. All this mesmerizing biological functionality simply on an intellectually perspicacious persuasive level states that a living cell in the past never arose from nothing, it is not happening today, and the mathematical probability of it ever happening in the future states, *it is never going to happen!*

For example, thinking and non thinking evolutionists casually claim that reptiles over time, millions! billions! trillions of years ago changed into birds! 'Evidence presented usually draws attention to skeletal similarities or some other minor changes/adaptations. First, fully formed reptiles which only produce fully formed reptiles are used as a starting point for this delusion. What is completely overlooked in this proposal are the following for reproductive reptilian regular results.

An Existing

Galaxy

Sun

Earth

Water

Oxygen

Food

Mates

Apart from the above there are tens of 1,000s of other interconnected things necessary for a male reptile to reproduce, not least being a perfectly functioning and biologically matching female. The specious argument next maintains that 'minor tiny changes' occasionally occur—this is true, but all those are simply built in adaptive mechanisms to accommodate variations in the atmosphere or environment. Beaks on birds—feathers on pigeons—human or animal skin colouration ad infinitum are all simply beneficial variations, chemically choreographed chromosomes flawlessly transmitting hereditary information 'silently and insensibly' to the unaware and unappreciative recipient. Recently Channel 4 TV in the UK, broadcast "The Great Sperm Race." It was the visually stunning story of conception. Words on paper cannot grasp all the molecular daily marvels of sperm production or monthly wonders of the ovum. So many facts, so many mysteries—so much knowledge leading to so much ignorance. Such joy. Such sadness. Such pleasure. Such madness. The sperm 70 days (approx) in production/storage—the eggs, 15 years (approx) waiting, longing for release! From 5 hundred million sperm, lovingly and pleasurably blasted off for an incredibly long and difficult journey, hunting, searching for just one thing, that egg mysteriously pushed out of the ovum, waiting to be fertilized in the fallopian tubes. It happens—hourly—daily—monthly–yearly—ceaselessly—flawlessly—beautifully—miraculously—wondrously—tirelessly and ecstatically!

Sperm penetrating egg.

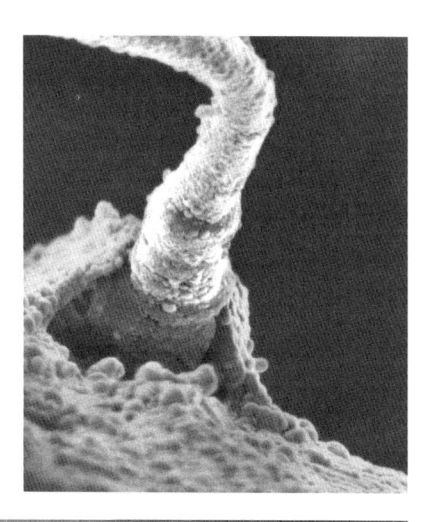

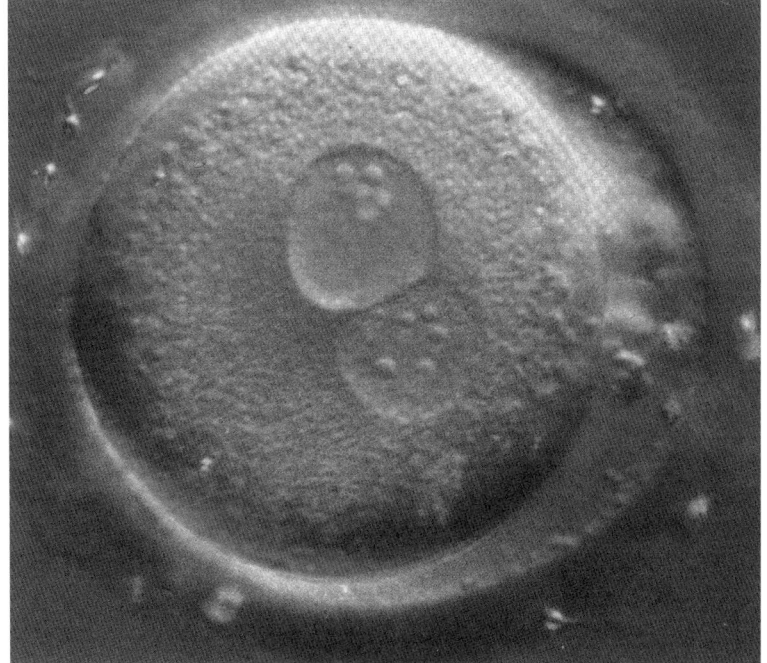

The exact moment of fusion!

Pictures taken from; Lennart Nilsson's book; *A Child Is Born.*

THE ORIGIN OF SPECIOUS NONSENSE

CONCEPTION TO BABY

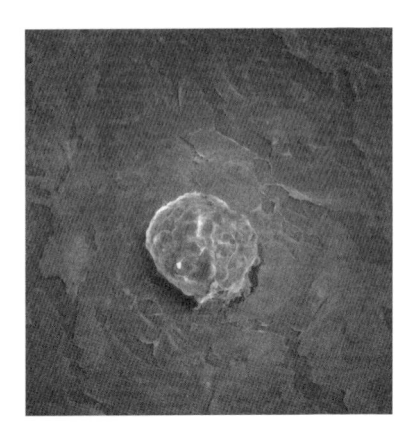

Picture of 5 -28 day embryo!

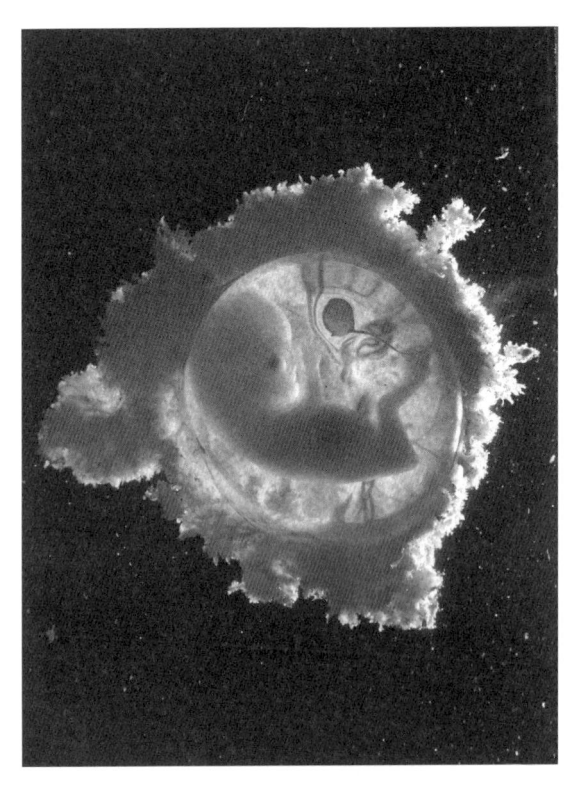

Beautiful baby in womb!

ONTOGENY

SPERM

EGG

DIVISION OF CELLS

BLASTOCYST

EMBRYO

FETUS

BABY

The above outline is 'concrescence' the growing together of related parts tissues or cells, the amassing of physical particles. From tiny, virtually impossible to distinguish cellular certified codes, emerge in time, teeth and they are not in our anus! Eyes not under our feet! Legs not growing out our necks! Noses not on bellybuttons! Tongues not under arms! Brains not in stomachs! Ears not stuck on foreheads!

Thank goodness, about 99.9% of the time there is physiological perfection and the reasons for that are 3 billion plus chemical instructions which were invisible throughout history but are now observable in the human genome!

Not all evolutionists are atheists (bizarrely!) but all atheists by definition are evolutionists! What both share in common (apart from delusional faith) is serious lack of scientific information necessary to substantiate their image of maggots to slugs to fish to reptiles to birds, animals and man! For example to develop up the biological ladder from 1 x cell to 1 x human requires staggering amounts of new genetic information! New DNA codes are essential for manufacture of skin, eyes, nerves, bones, ears, muscles, blood cells and a myriad of other physiological attributes. Mutations cause a net loss of information.

The essential difference between a human and a bacterium is in the genome/DNA all other differences follow on from that. Saying that x amount of DNA in apes (or chickens) is similar to man's is pointless because there are 12 x million 'chemical words' arranged in intelligent sentences which are definitely not similar, that is 1000 x 500 page books and it is this specific genetic code or formula that forms us and makes the human race 'kings of creation.' Some 12 x million chemical unique words in 12 x million (or billion) years is a biological chemical genetic impossibility to change of its own violation from animal codes to impossibly precise human codes . . . *it simply does not happen and is scientifically not credible.*

Charles Darwin wrote in his 6th edition of what he viewed as his 'execrable book.' "Natural selection is incomplete to account for the incipient stages of useful structures" this is correct, for half an animal does not work. The well known evolutionary biologist, Steven J. Gould said, "half a jaw or half a wing is useless." To propose that mutations in tandem with blind selection are the root causes for over 1,000,000 viable enormously complex species is to mock logic—deny the weight of evidence, reject mathematical probability and ignore perfect biological precision.

The eye has: 1. Automatic aiming 2. Automatic focusing 3. Automatic aperture and are extremely attractive exactly where they are positioned in our faces. There are in the region of 7 billion pairs of human eyes on the planet and all look different in distinctive shaped faces! How is this possible? That's like 7 billions sets of coloured marbles all stuck in the face of 7 billion snowmen and all looking individually different! I don't think so! They would all look exactly the same. This is the genius of our eyes that even the honest, solitudinarian, Charles Darwin was led to express about the formation of our wondrous eyes. "It is absurd in the highest degree to suggest that the eyes could have formed by natural selection." Vision is vital for survival, every component is necessary. Half an eye is useless. Therefore, if by his own admission the eye cannot create itself, then it is strange that "he could not see" the scientific impossibility of random accidents, destructive mutations and blind chance over time being the cause of his fictitious dream—evolution. At least the man did not try and hide his inconsistencies as some of his modern deluded followers have done and are doing. His book has the colossal total of 1515 suppositions and I have counted every single one personally.

Biological evolution is actually an oxymoron. *Bio is from the Greek word 'Bios' meaning 'life,' and evolution in the atheistic sense means 'Death,' since non living* matter is actually dead or lifeless and they believe that somehow! Someway! Sometime! Something! In someplace! happily happened through chance that brought life into being from death! This suspiciously sounds to me like some form of resurrection!

I know some people seriously postulate that a species of modern birds are descendants from lizards! (Imagine!) Others say 'flying squirrels used to fly, (or is it maybe one day they will be able to fly?) No matter in genetic genius they see only what they wish to see. They view all creation as mere lucky trillions of accidents and dismiss the following quote and prefer to believe the unbelievable, "The probability of life originating from accident is realistically comparable to the probability of the unabridged dictionary resulting from an explosion in a printing factory." This revealing amusing statement is from Dr. Edwin Conklin, Princeton professor of biology, as written in Readers Digest, January 1963, under 'The Evolution Revolution.' Another astonishing admission is by Dr. George Wald, professor emeritus of biology at Harvard and Nobel Prize winner in 1971 who wrote honestly, "There are only two possible explanations as to how life arose: Spontaneous generation arising to evolution or a supernatural creative act of God . . . There is no other possibility. Spontaneous generation was disproved by Louis Pasteur 150 years ago—but that just leaves us with one other possibility . . . That life came as a supernatural act of creation by God, but I can't accept that philosophy because I do not want to believe in God." From: *Origin, Life and Evolution; Scientific American* (1978)

There it is in a nutshell. People believe what they want to believe, regardless of scientific evidence. It's obvious the explanation value of the evolutionary hypothesis of common origin, is zero. Millions have been misled about scientific support for this theory to the point that many, many scientists have swallowed this bluffer's bunkum and regularly regurgitate it for public 'con'—consumption!

If people would stop and think about the lunacy of the propositions being advanced and the realisation dawned, there is no—none—zero- evidence for evolution, then what I am about to write and you are about to read will surely make you realise it is the greatest hoax and deceit in the entire history of science.

Ontogeny is the study of an individual organism from conception to birth. However, a number of biological facts are beneficial for good mental health when one is confronted by specious spurious splutterings on this toxic topic.

For a man and woman to produce a healthy baby through love the following seven biological entities are crucial:

A perfectly functioning penis

A perfectly functioning vagina

Mature sperm

One ripe egg

A healthy male and female body

A healthy womb and placenta

Ongoing healthy pregnancy to birth

Now please stop and ask yourself; Will half a penis do? Will half a vagina suffice? Half a womb? Half an egg? Half a placenta? No, no, no, a million times NO!!! There are tens of thousands of women across the world who cry regularly because there is some tiny flaw in their reproductive organs (or their partners). At the heart of the unscientific theory of evolution is reproduction. For species to be alive they had to reproduce. For reproduction to have taken place the 7 biological necessities mentioned are critical. Remove one and disaster shall surely strike. Pregnancy is not a game of chance like poker, it is based on physiological certainties and biological bullets all firing at predetermined times to produce the magical mysterious miracle called life. In humans this tiny boy or girl when born is, somehow, a reflection of us. This helpless little person fills our hearts with pride and inexpressible joy. They are perfect, we see no flaws, we are consumed with feelings of power and helplessness yet also with passionate parental protectiveness. As sentient cognitive beings we instinctively sense at times like these a force—being—spirit—God—creator, architect or whatever, greater than ourselves! And depending to a very large extent on our backgrounds, those thoughts and questions remain, just so! The one sure thing many of us are persuaded of is: all this biological beauty—chemical clarity—organic growth—hopes—dreams—visions—goals for ourselves and our loved ones have some ultimate meaning, point, purpose and possibly a limitless and timeless peaceful future! Many Jewish prophets spoke of a future government that would establish peace and harmony worldwide one day!

I don't care what religion those people were, the point is they appear to offer our struggling, suffering, searching, seething society genuine hope for a better life in a better world.

The current ongoing conflict continues between;

Theory V Reason

Faith V Facts

Evolution V Reality

Tradition V Science

Belief V Truth

Speculation V Logic

Suppositions V Evidence

Theory—evolution—tradition—speculation—suppositions, ad infinitum is the language of despair! It is extraordinary that millions of educated good people believe the unbelievable, and advance, as if in a trance, the un-provable. Simple questions are powerful, for example: The following profound question was asked by British astronomer, Paul Davies; "The greatest puzzle is where and how all the order in the universe came from originally?" Albert Einstein not in answer but in harmony pondered, "The high degree of order was somewhat of a miracle. In fact everywhere we look in the universe from the far flung galaxies to the deepest recesses of the atom, we encounter order."

Theoretical probability is not provable science. Without observation to check theory, at what point does maths and science become speculative game playing? Order is everywhere, even in the seeming chaos of galactic explosions, volcano's violent storms, there appears to be seeds of orderly deeds! Einstein reminded us that, what we don't know far exceeds what we do know, and if only we could heed Christopher Hitchen's wise words to remain silent on topics we are ignorant of!

Being born in a barn does not make one a cow, and not having a degree in science does not make one a fool! People who have science degrees are lucky as they have a window into worlds of wonder that most humans are ignorant of. Many graduates' personalities are formed by the time they set foot in their first job. What some do not realise is there are factors of bias through background which play a significant role in how they view the world and why acceptance of various theories and ideas appear normal to them. Very many young people have three eyes! Innocence—Ignorance—Idealism. I possess to some degree all three. However three things modify and protect us from becoming slaves to others ideas, opinions and theories they are, Knowledge—Wisdom—Common sense. Knowledge particularly applies to facts or ideas acquired by study, investigation, observation or experience. Its opposite is ignorance—the most expensive commodity on planet earth! Wisdom means having understanding of people and situations and unusual discernment in dealing with them. Its opposite is 'simple.'

Common sense means having the ability to reach intelligent conclusions and suggests an average degree of such ability often with native shrewdness but without sophistication or special knowledge.

The sad thing about common sense is that it is particularly uncommon. It protects when it comes to basic dangers such as speed, fire, gravity, the ocean, among other things, but appears to abandon us in our pursuits of pleasure and reason.

When we are propositioned with ideas or theories for the first time most of us listen as it is a sign of respect and good manners. If our curiosity is aroused we may seek further information and without realising it, depending on background be sucked into a way of life that could split family and friendships for years. How then do we protect ourselves? One way might be to seek out and read about happy contented individuals and find out the source of their peace. The following are seven characteristics held in common by contented people regardless of gender, race, culture or religion:

Love of people

Tolerance of differences

Spirituality

Purpose

Being loved

Good friends

Loving family

Our enjoyment and contribution to those seven vital characteristics is indicative of a person who wisely appreciates true values.

As I read regularly excellent articles in Scientific American, American Mind, American Scientist, National Geographic, New Scientist, plus many other magazines, I ask myself a simple question: What planet are those writers on? I can state without fear of successful contradiction that when they pontificate regarding origins and evolution they simply do not know what they are talking about. And worse, they use arcane, esoteric language to befuddle what is already muddled mysterious musings on origins. A rudimentary knowledge of DNA, reproductive sex, ontogeny, genes, chemistry and biology catapults the mind from the shameful constraints of evolution in general and religion in particular. There are not three choices, there are only two. Either there is a complex powerful mastermind who created everything or everything is the result of nothing or chance, with no mind the master of anything! Atheists are honest; they state openly their personal construct: There is no God! And most evolutionists agree. However there are some evolutionists who degrade the concept of a Cognitive Artistic Genetic Engineer (God) with the chemically impossible notion of spontaneous generation leading to natural selection. Let's get this straight. Darwin's tree of life is laughable in the light of DNA sequencing. Natural selection under the glare of modern microscopes collapses because it cannot be observed occurring. Once the penny drops that as a proposed mechanism for biological molecular genetic DNA reproduction, it is scientifically impossible and is in fact a hoax, then the evolutionary rose coloured glasses fall off!

We don't need quote after quote, fact after fact, point after point, proof after proof and why not? Because when natural selection, evolution, is exposed as a scam supported mostly by atheists who are as demented in their delusional convictions as are the religious leaders of most organised religious persuasions, then we can embrace reason and logic with all the passion of an honest philosopher, hungering, seeking and searching for ultimate meaning.

Natural selection is the unsupported, unscientific, untruthful, unedifying, unbelievable teaching that chance chose camels that once were maggots. That chance chose beautiful bears which were once whales! (or visa versa—who cares both are mind bogglingly stupid) that chance chose elephants who once were fish! That chance chose dinosaur's which once were rats! That chance chose eagles that once were frogs! That chance chose horses that once were dogs! That chance chose crocodiles that once were snakes! That chance chose swans that once were chickens! That chance chose butterflies that once were crows! That chance chose kangaroos that once were bees! That chance chose gorillas that once were salmon! That chance chose sharks that once were trout! That chance chose man that once were monkeys! Silly yes! Illogical yes! Unscientific yes! Ludicrous yes! And yet, yet as I write this in the quietness of my home and as you now read the words I have just written, millions at this moment believe such gibberish . . . Oh, yes, the erudite educated shall rush to condemn what I have written (and me!) but I couldn't care less! I say stop pretending there is proof for the greatest deceit in the history of science. Produce the facts or be quiet. It is a challenge the honest shall wrestle with, the deluded will fight, the wise may ignore and the confused might think about. I repeat natural selection, i.e., the selection of bodily improvements by chance is not happening now—never did in the past and with all the available DNA and scientific knowledge, we know for a fact is never going to happen in the future. Why? because of biological, extremely extraordinary, excellent genetic laws. All over the world, evidence shows today and yesterday that species produce after their kinds. It's as simple, obvious and as powerful as that.

In January 2009 edition of New Scientist, the front cover courageously showed a tree and imprinted across it were three truthful words: *"DARWIN WAS WRONG."* I will not gloat or traduce the educated intelligent owners and writers of this impressive magazine. I shall not point to quotes (out of context) to support my certainties from the article. The editor in his wise commentary quotes Michael Rose of the University of California, an eminent evolutionary biologist, who said; "The complexity of biology is comparable to quantum mechanics." This is a lucid, important and honest comment that will be agreed upon by enemies of both persuasions. Why? Because quantum theory/mechanics and biology are fundamentally so mind bogglingly complex that portentous pretentious pronouncements are easily exposed!

There are scientific fixed laws that marry easily to fixed scientific facts. Evolution is assuredly not one of them. Richard Feyman as quoted in John Polkinghorne's book on Quantum Theory said. "I think I can safely say that no one understands quantum mechanics." In his introduction Chris Isham of Imperial College London wrote: "John Polkinghorne has produced an excellent piece of work . . . Many authors of 'popular' books on modern physics have the regrettable habit of mixing science fact with science fiction. Polkinghorne never does that: he always *allows the truth to stand by itself* and show its own fascination . . . I think that this is an excellent contribution to the literature on quantum theory for a general audience."

Is it not a remarkable admission that "Many authors of popular books on modern physics have the regrettable habit of mixing science fact with science fiction" to strengthen their ideas? In the same issue of New Scientist, in 'Opinion Essay,' I quote the following, "Its easy to be seduced by the nature of logical thinking and its illusion of certainty!" Hello! What? Yes that quote might be out of context, but no matter, for to ordinary sensible individuals reading it, aspersions are cast on an eminently wise cognitive procedure: 'Logical thinking.'

In the real world, this silly comment qualifies as assuredly assertively asinine. If we go to a doctor or surgeon with some ailment, we want him/her to think logically and be absolutely certain his educated prognosis is accurate and not by any stretch of imagination an ignorant ignoble destructive detrimental illogical illusion.

DNA is the true scientific mechanism by which we reproduce ourselves. No sperm, no life. No seed, no fruit. No DNA, nothing. And no words written or spoken by anyone can invalidate that scientific fact. DNA was us, is us, became us. Seven short generations ago there was not a biological trace of any one of us and in seven generations to come we most assuredly shall not be here. We are chemically three billion letters held together by hydrogen bonds that provide stability for Adenine—Cystosine—. Thyanine and Guanine, the four vital microscopic elements that helped make me and you who we are. DNA is not random, it appears random as it transforms with military precision into a blob of purposeful flesh and blood called a blastocyst. Ontogeny the development of a baby from the moment of conception, is not haphazard and silently, mysteriously, beautifully, scientifically, obey immutable fixed biological biochemical biogenetic laws. The next 280 days (average length of pregnancy) not thousands—not millions—not billions—but yes trillions of chemical cellular actions grow intelligently inside a living womb; and 99.9% of the time produces a beautiful healthy baby.

THE MOMENT OF BIRTH
Now comes the climax of a true miracle, the crescendo of chromosomal complexity, the heartbeat of all beating hearts, the power of a tiny life to live, the pushing of a mother's love in pain, the struggle to survive in its true setting of blood, sweat, tears and years of preparation.

The baby is having a hard time too. There is a surge in production of adrenaline (marvellous!) and noradrenalin the stress hormones, to levels higher than at any time in later life! (Coincidence? I don't think so!) Now pain, joy, excitement are mingled. The experience of pain varies from one woman to another, depending of sensitivity, expectations and anaesthetic effects.

THE SUDDEN SWITCH
For the baby, birth itself is a cataclysmic event. The adrenaline shock counteracts the oxygen deficiency in the final stage and

prepares the baby for *the sudden switch to breathing through their lungs.*

At last, out in the real world—a world of dazzling light, cold air and loud noises. Almost immediately comes the first cry—a rare, sometimes hesitating sound.

Now 25 million little air sacs (alveoli) must be filled with air. Up to now, they have held fluid, but this is rapidly expelled in blood and lymph. (Lymph is a clear watery fluid that contains white blood cells and circulates throughout the lymphatic system, removing bacteria and certain proteins from body tissues, transporting fat from the small intestine, and supplying mature lymphocytes to the blood.) The first breaths are among the most arduous of one's whole life. The bloodstream must now be redirected. *The hole in the partition between the heart's atria is sealed.* In a tiny babies heart is a powerful fibrous muscular bridge that literally saved the babies life during pregnancy. It is called, "The Ductus Arteriosis."

Because our lungs don't work before birth, this vital crucial amazing 'bridge' shunted blood meant for the lungs into the aorta, the bodies biggest blood vessel which arches out of the heart and down towards the lower body. It has an internal diameter of about 25mm (1 inch) and blood gushes through it at a speed of about 20cm (8 inches) per second. At the very moment of birth, this hole is sealed! *It seals itself!* It must for baby to take its first ferocious breaths and the crucial liquid of life now for the first time flows freely and beautifully throughout the newly born babies miles of perfectly formed veins. Astonishing. Astounding. Amazing. Yes! Mutations. Chance. Randomness. NEVER.

Up to now oxygen has come from the mother via the umbilical cord, but now the baby is self-sufficient: the blood must be directed into the lungs and then all over the body. For the parents, this is a miraculous moment. Such a tiny human being, so full of life—their very own child! The astonishing intelligence and design is quite literally 'breathtaking.'

Language, words, poetry, songs, gifts, laughter, looks, are what we use to convey ideas and love. The conception to birth of a baby is filled with so many so amazing biological personalised features (all in favour of baby) that any other conclusion other than, an original originator of organising staggering complexity is quite simply laughably ludicrous and in reality not credible.

Evolution asks us to believe every child born every minute of every day in every country is the blind result of random chance and destructive mutations! Asks us to believe every animal of every species came from nothing! Asks us to believe every type of fish and other amphibians which themselves came from slime or ancient soup! Asks us to believe every single beautiful flower designed itself and their lovely scents. Asks us to believe bananas manufactured their own skins, oh. and by accident the fruit inside, oh. and by chance the stalk, oh and through random chemical chance the brilliant minerals so beneficial for our skin and health. Oh, and through another accident threw in the amazing vitamins so necessary for nutrition. Oh, and by the way through more random mutations by chance the seeds that helps more bananas grow! Asks us to then accept not just banannas but every single fruit with every single specific seed in every single country is also the result of chemical chance, rancid randomness and myriads of myopic mutable mutations! Asks us to believe every single food that enters our mouths, every single breath of free oxygen, every object of amazing beauty seen with out eyes, love—stars—planets—universe—cells—everything—everything—everything came from nothing, nothing, nothing! It is simply asking too much . . .

Now I ask and it is time to ask—why do intelligent people choose to ignore the obvious evidence of their eyes? Feelings of their hearts? Logic in their brains? A blind person can feel an object and reveal what it is and who made it. A deaf person who has never heard music can still dance with friends at a wedding because he/she gets the sense of joy and celebration. A child of atheists is confused by the promotion of meaningless nothingness. Like children of religionists frightened by *hellfire* and appalled by the idea that God randomly kills people! (i.e., God took them!)

People think and reason differently when told they are dying, and have less than a year to live! The obvious reason being life is precious and death obnoxious.

All the philosophic pills, religious rantings, political problems, family feuds, friends foibles, children's childishness and future plans utterly, suddenly, dramatically and irrevocably change! In one bizarre way they change for the good in the sense that common sense overcomes nonsense! Now once again *childlike* reason returns in the season of endings! The deathbed is breath-led and avoids self deception, and the mind becomes like an aerial that receives reception of reason based on intelligent cogent conclusions.

Subterfuge like fairy tales are banished and hope no matter how slim or unlikely is welcomed. Days of wine and roses are fondly remembered and unquenchable longings for life anchor the soul to possibilities! This is no time for phantasmagorical theories that offer bleak blackness as a future! Gone is the passion for ideas in fashion! Gone is the pretence that reason was the guiding principle! Gone, forever, is the fiction that life is a pointless puzzle. And last but not least the extraordinary disappearance of any animosity and antipathy one may have felt towards the source of life, God!

The mind is a universe of complexity, an instrument of government or a tool of destruction. The facts are it is generally only capable of reacting to what is fed into it and that is why so many are sick! The mind's medicine is compassion—reason—logic—hope—education—love—joy—purpose—health—friends and family. The mind's poison is: evolution—organised religion—tribalism—nationalism—racism—hatred—violence—lies and selfishness.

The body may be dying but the mind is living and it is this that speaks silently of purpose! There are only two choices of interest; either there is a munificent marvellous manufacturer—God, or there isn't! And the dying heart through reason and possibly desperation now wants to believe there is! Is it knowledge, selfishness, fear or understanding that leads thoughtful people to one simple conclusion? "I know in my heart there is something or someone out there." That realisation does not make it true and neither would its opposite make it false! Ignorance of God's reality does not make him unreal! Not knowing his purpose doesn't mean he has no purpose! Being tormented over mankind's sorrowful suffering shameful situation does not mean there is no satisfactory explanation!

Evolution cannot and does not give any hope to the living, dying or dead! Organised religions are diseased disorganised disasters, dastardly dishonouring God with daily drugs and their dreary delusional depictions of despicable dyspeptic dogmas! So is there any hope? Yes, and it is to be found in the unlocking of believable origins as outlined in a Jewish 3500 year old book called Genesis. It reveals the planning, point, purpose, and **WHY?** The Cognitive Artistic Genetic Engineer (God) bothered!

Let Jews feel morally superior and special!

Let Muslims pity them!

Let Christians hate them!

Let atheists continue their indifference!

Let agnostics smile benevolently!

Let the non religious enjoy freedom!

Let the rest be at best—confused!

However, the fact is there is an unbridgeable chasm between gentiles and Jews; reason being, revealed origins—laws—and purpose to that lucky family wandering around the desert in the middle east as they herded animals and lived their lives in tents! The most famous Jewish man who ever lived quoted from Genesis many times . . .

To those dying I say read it with an open mind and to those living I say, what have you to lose by checking out this ancient scientific record of mankind's beginnings! If it is true you have gained, and if it is false you have lost nothing and possibly increased historical knowledge and curiosity.

The certainty of ignorance camouflages and distorts scientific facts! For example many people have accepted evolution on the basis of "scientists have proved" "scientists have discovered evidence" and yet all my life I have offered 1000 pounds or dollars depending which country I was in, for any person to prove to me unambiguously that on a molecular level one species changed into another. In fact it comes as a shock to most people when a non religious person like myself points out to the deluded there are no scientific facts whatsoever to support this laughable ludicrous toxic theory!

I personally care nothing for what people think, but I am outraged by educated scientists who pretend, yes pretend, there is evidence to support this fairytale. The Empire State Building cannot be supported on a Styrofoam coffee cup from Dunkin Donuts; nor a rocket go to the moon on a litre of petrol from Shell; nor the sea fit into a child's bucket on a Californian beach and yet in mathematics there is more chance of all three happening (which never will) yes all three occurring than for the awful soul destroying fiction of evolution to be true. The frog never changed into a prince, the gods of Egypt are fantasies, Dracula never existed, the king was naked and evolution is a demonstrable hoax just as surly as Santa Claus.

Yes I know in spite of my book, in spite of facts, in spite of reason, in spite of reading the courageous statements of many scientists that: "Evolution is a hoax, a joke—a fiction, a cancer on the mind of an un-provable, un-demonstrable unscientific fraud." I know that many will still choose to believe it! I don't care. Only don't ever try fool me with specious arguments.

I now address myself to scientists who proclaim evolution as fact. I shall on a biochemical level expose you as frauds. The damage you have done to innocent minds and hearts is incalculable. The priceless hope for a secure future of peace you have devastated. The social carnage of the 20th century can surely be placed at the door of evolution and her *"high priests of the highly improbable foisting the impossible on the impressionable!"* Evolution is intellectual defecation, mental pollution, cognitive dissonance; it is cocaine for some in the thinking classes to give a false high for a true low! As a theory it has nothing to recommend it, nothing to support it, nothing to be proud of. It is a disease within society just as surely as organised religions.

The fact that some eminent luminaries in the scientific fraternity speak glowingly of it has as much relevance as the Nazis promoting the obnoxious fiction of Eugenics or an alcoholic promoting the felicitous feeling of drunkenness! All are specious and have social personal and moral consequences.

FANTASY
Ramesses I, shown here being greeted by gods in the afterlife; this fiction was promulgated by ignorant priests and believed from the greatest to the lowest caste in Egyptian culture.
Picture taken from: The USBORNE, Internet—Linked Encyclopaedia of Ancient Egypt.

Enemies of Reason

Organised Religion

Evolution

Childhood Teachings (some)

Social Influences (some)

Traditions (some)

Tribal Loyalties

Fear

Friends of Reason

Thinking

Logic

Evidence

Spirituality

Parts of the book of Genesis

Acknowledgement of God

Limitations of time—science

The human mind cannot grasp a time before time began! Space before space existed! An endless universe that ends in endlessness! The creation of matter from nothing! A time when nothing existed! A God who has no God! A Creator with no Creator! A designer with no designer! To apply the term "Infinite Regress" to those concepts is churlishly childish since its actually an oxymoron because infinite by definition cannot be constrained by something that moves back from a starting point! Because the mind cannot grasp the concept, a ridiculous smokescreen is lobbed into the argument by obscurants to obfuscate complicated metaphysical musings . . .

People in most modern civilised countries are free to believe the unbelievable and free to reject observable reality and reason if they so wish. All of us to one degree or another are pre-programmed by time, place and background to accept or reject certain ideas. Young people because of innocence, lack of experience and idealism are regularly trapped into ideologies which suck them of independence, courage, originality and leaves a shell/zombie to carry out the instructions and wishes of the "inspired" leaderships! Tragic but true! No adult in their right mind could be convinced Santa Claus is real!

Yet a similar fiction has infected the minds and hearts of adults since 1859 in particular when a hoax of phantasmagorical proportions exploded in western culture. It was due to a life long study of insects—birds—(pigeons in particular) vegetation and animals by Charles Darwin and culminated in his laughable notion that animals and fishes, flowers and man and well, everything, came from something else, that's why, all we see around us, well, just is! He called his fable *On Origin of Species* which I have forensically read and counted 1515 (yes one thousand five hundred and fifteen **SUPPOSITIONS.**) That's about four for every day of the year for four years! I defy any disciple of this deluded, tortured, nice man to explicitly and unambiguously point out to me on any of the laughable 384 pages a shred of rock solid scientific proof for his delusional drivel in relation to provable origins!

I challenge every single Darwinian evolutionist on the planet to fact facts and wake up and smell the coffee! Organised religion has originally deluded most of you—evolution half of you and scientific gobbledegook and psychobabble the rest.

I don't mean to be offensive. I mean to shock the ones among you that are amenable to reason. Evolution is a non-provable theory! Forget the stupid Scopes 'Monkey trial' and see those religious folk at that time for what they were—poor, southern, frightened, confused, bewildered, brainwashed racist pseudo Jewish Christians! Scopes had the right constitutionally to believe what he wanted and possibly the legal right to teach it. However in the hysteria what was not pointed out to him was this: it is unscientific to teach fantasy as fact and if there are no facts, then this idea should be discussed in a philosophy class and not in a science one.

There are, in particular, two specialities of mine that on a scientific and biochemical level demolish the fictitious foolish fantasy commonly known as Darwin's theory: Evolution as is generally understood means we all came from something other than our own species and they came from nothing! Some shall deny this and try to explain something else they also do not understand! Once one steps outside the safe parameters of provable science, we are into science fiction!

My two disciplines are:
    A human living cell
    Ontogeny—(from conception to birth)

How do I express the staggering chemical complexity of this cellular universe of astounding interrelated interesting magical mystery? I can't! I can only say a cell is a world within a world of amazing molecules/proteins/atoms/enzymes all chemically working to keep us alive and healthy. It could not have designed itself, and we have trillions of them all working not by chance but by mysterious laws every second of our lives. Why does my mind see instantly and profoundly with absolutely no ambiguity whatsoever the result of reason and reject the fantasy of chance? That is why, I suppose, evolution is surely one of the few things on this earth that enrages me! (War and child abuse being two others oh yes and a third—organised religion . . .)

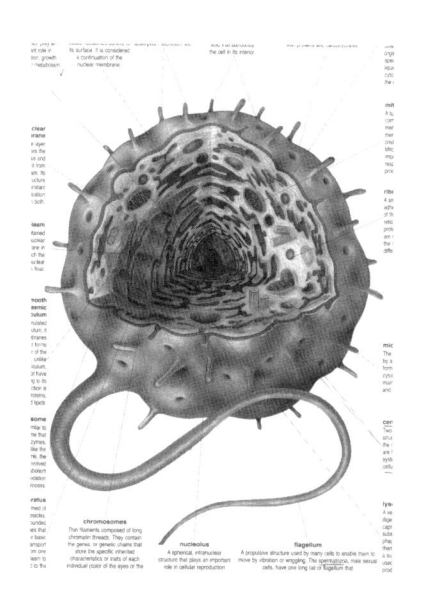

From one cell to a baby

REALITY

There are few atheists in a foxhole! Fear of dying is part of living. A happy person loves life and no one enjoying it wishes to die. The old are young in their hearts. The young cannot imagine being old. A young happy mother or father on being told they have inoperable cancer and shall shortly die (like Jane Goody recently in the U K) are utterly devastated at its random unfairness. Liam Neeson (Irish actor) recently lost his beautiful wife Natasha Richardson in a tragic skiing accident and his world collapsed. Death is unnatural unreasonable unreal unwelcome and unlikely to bring joy or happiness. Life is a priceless gift and loosing it an outrage.

Evolution's depressing explanation: There is no God—no point; no purpose; no future; no hope, therefore, get all you can while you can! The worst thing of all here, in my opinion, is the 'no hope,' but they will retort; "We would rather have no hope than false hope" (me too!) The problem for them is they see 'hope' as expounded by various religious cults as fantasies concocted by priests to keep their followers placated and duped. And I have to agree as most teachings of religions on this topic are farcical! Atheists are fearlessly correct on where the dead go and what happens to them. We all know in our heart of hearts that we don't like the idea of being put in a box of wood and placed in a hole of cold clay. Is there a realistic alternative at present? No. Is there a realistic hope for the future? Yes. "Faith without evidence, as Sam Harris cogently writes in his excellent book, *The End of Faith,* disgraces those who claim it!" However, I choose hope over the word faith for a very good reason: Faith is a religious term but hope universal . . .

I have hope that the supreme intelligence behind order in the universe has a purpose. I am intellectually interested in him as God or the Cognitive Artistic Genetic Engineer! I have read all the major philosophers plus the well known religious books. None of them stand up to forensic analysis against common sense, except one, ancient Hebrew and Greek literature. I accept some and reject much of it; the point being I will read and evaluate anything that I am told holds out hope for suffering humanity. And why not? The past 100 years alone has witnessed over 100 million lives smashed, ruined, brutalised, raped, burned, bombed, killed, wrecked and murdered by madmen! We are divided by books, ideas, words, memories, traditions, tribal loyalties, religion, and politics. Once evolution is thrown into the mix, its toxic amoral animus animates animal instincts in some to seek and destroy what are perceived to be weaker species deserving death. Their sickening opportunistic idea of themselves as a superior species struggling to live as the nominated 'survival of the fittest' is the fuel that fires this dreadful drivel with such disastrous consequences. No man, no ideology and especially no religion can unite the human family. The tragedy is we are all so similar in our dreams, desires and ultimate destinies. As we sleep-walk towards catastrophe, it is only a matter of time before weapons of massive destruction are unleashed on our planet! The population explosion alone with its gargantuan appetites for water, oil and food guarantee it. In view of this rather pessimistic viewpoint (based on past violent experience throughout history) is there any hope for mankind?

The answer is yes, based on;

A Creator

His promises

His purpose

His solemn word

His kingdom (government)

His original idea of eternal youth for mankind

His Messiah or chosen King of his future earth wide peaceful rule.

There is no other realistic hope as everything else is pie in the sky. Mock and laugh if you will, that's understandable and ok! But I challenge anyone to come up with a more reasonable, responsible, realistic, believable workable reality. We were created to live not die, and every thinking person during their last days would want it—wish it and hope it to be true. If there's a God, it is, if no God, no chance . . . you choose, I know I have. . . .

EVOLUTION V REASON

Many who promote this idea do so because it is a cause! It comes as a shock to be told: there are absolutely no facts to support this theory. None . . .

A good argument starts off with a sound premise. A bad argument invariably obscures the fundamentals with specious notions. For example, the personable, strident, likeable Richard Dawkins, et al, will dismiss offhandedly many religious leaders (as I do!)and refuse to give them an ear! However, this is a mistake because even if a liar tells you the truth it is still the truth! Some scientists, anthropologists, palaeontologists, biologists, geneticists and a lot more besides are fond of one particular idea and it is, *time*. It is patently false in the manner in which *time* is discussed. Whenever they come across some new/old discovery, for illustrative purpose, e.g. a bone, we are confidently told it is 500,000,000 years old (or 20 or 70!) The point is; they do not really know because no man or woman living can be truly confident when they discuss 500,000 years ago (and in the geological scale of time that would hardly qualify as a blink!). Never mind 500 million years ago! And if my book wants to nail home one vital point in my attempt to expose Darwinian evolution as a joke and a hoax it is this: read between the lines and remember, evolution is simply a game called 'guess' and the language is always as follows:

Maybe

Possibly

Perhaps

Probably

If

Presumably

Appears to be

Apparently

Assumption

Conceivable

Speculate

Chance

Likely

Might

Imagine

May have

Could be

Presume

Opinion

Perhaps

Always keep in mind it is just a theory among thousands, *I just reject one more.* So little evidence for such a proposition is quite ludicrous. It is truly in today's frenetic world the best example of the worst form of wishful thinking. Its falsity is part of its cultural poison for which the cure is reason and common sense. It is very easy, in fact, to demolish the so called argument for evolution once we realise the proponents have, no facts, no science, no idea, no knowledge of how life began, are guessing, are bluffing and disagree among themselves. As Scott Adams lucidly pointed out in his book, *God's Debris,* "Evolution isn't a cause of anything it's an observation, a way of putting things in categories. Evolution says nothing about causes."

Atheistic evolutionists never tire of claiming that people who are intellectually convinced of a first cause—God—are deluded and devoid of reason. I state emphatically that the opposite actually is the factual case. The American Heritage College Dictionary—(Third Edition) states under Reason; "An underlying fact"—"The capacity for logical, rational and analytic thought; intelligence"—"Good judgement, sound sense."—"A normal mental state; sanity."—"To use the faculty of reason; Think logically."

Children are logical thinkers. They constantly ask, why? all day, most days. Reason appeals to them and illogicalness repels them. For example a child will one day surely ask the age old question: "Mam, Dad, where did I come from?" As a child I was told: "A stork delivered you to us." I wondered: "Where did the stork get the baby?" . . ."Why did the stork bring it to our house?" . . ."Where did the stork come from?" . . ."Did the stork have babies too?"

A child is cognitively hardwired to speak truth and ask straightforward, powerful, simple questions. They expect to be told the truth and in a home where honesty is seriously promoted and respected that child does not know how to lie. It is only when she/he hears and witnesses deceit that it creeps into their minds and vocabulary.

The most powerful idea in history (for good and bad, unfortunately) is the conviction of a first cause: i.e. God. The most pernicious idea of the past 200 years in particular is Evolution. The world of 1809 bears little resemblance to our world today of 2010. Back then there was no:

Internet

T.V.

Radio

Magazines (Rare)

Newspapers (few)

Cars—Buses

Airplanes—Helicopters

Electricity—Neon Ads

Central heating

Mobile phones

Oil (in the modern sense)

Gas

Lotto

Antiseptics

Modern hospitals

Flush toilets

Schooling for the poor

Variety in diet (rare)

Running water in homes (rare)

Modern medicine

Pharmaceuticals

Biros

Videos—CD's

Stereos—Records—Hi-Fi's

Refrigerators

Washing machines/Dryers

Computers—Facebook, Twitter

Telephones

Trains (very few)

Metal boats/Ocean liners

Modern police forces

Medical knowledge (extremely limited!)

Headache tablets

Post (by modern standards)

Credit cards

Government help for the poor and elderly

Supermarkets

Cheap—quick nutritious fast food

Ambulances

Unions

Cinemas

Celebrities (very few)

Football teams (few)

Sports entertainment (rare)

Workers' rights

Hospital births

Cows milk for babies (seldom)

Nappies (by modern standards)

Bars of chocolate

Rock concerts

Cameras

Toilet paper

Soap (rare)

Blades for shaving (in the modern sense)

Drugs (as available today)

Reading glasses (as available today)

Texting

Communications (as available today)

Pregnancy test kits

Typewriters

Atomic and Nuclear weapons

Tanks—Machine guns

Suicide bombers

Bad news on the hour

Dish washers

Tipp-ex

Calculators

Traffic

Christmas cards

Anaesthetics

Motorbikes

Dentists (modern)

Transplants

X-Rays

Nylon (toothbrushes)

Microscopes (modern)

International travel (on a mass scale)

Walkie-talkies

Torches (electrical)

Buggies (for babies)

Bicycles

Ice skates

Roller skates

Roller blades

Skis (modern)

Sunscreen

Sunglasses

Satellite dishes

Rockets

Petrol

Hamburgers

Microwaves

The 19th Century was a world of blinding grinding poverty for millions of suffering souls. Religion may have been of some comfort when people were at the end of their dreary days, but in so called Christian lands it was completely out of touch with reality! Suffering and anxiety cause humans to question and doubt their belief systems, this is not a bad thing! However, when the fantasy *Origin of Species* was first published in 1859 it was a magnet for educated minds within Christendom. They were discovering through science, travel and education that religions were a toxic parasite on humanity.

The poor on hearing about evolution looked around and said; "That's it, there must be no God to allow all this chaos, pain, wars and suffering!" So there it was. A scientific deceit, fable, hoax, scam, but it was more attractive to millions than the nonsense taught (and expected to be believed) by Christendom's ludicrous clergy.

Today we have all the mod-cons and none of the old convictions. Life's uncertainties, oftentimes bleakness and occasional joys, sometimes throw thinking people into a furnace of depression and firing line of doubt, questioning and confusion! No human being has all the answers, religion few, but evolution none. Why betray reason by adopting what eminent scientists have labelled, "A Hoax"—"Joke"—"The greatest deceit in the history of science"—"A fantasy"—"An impossibility"—"A fairy tale for grown-ups"- "A fiction." Therefore, every single time we hear about the "undeniable evidence" (so called) remember the above. Also if we are in the company of evolutionists inform them of Kent Hovind's offer in 2002 of $250,000 to anyone who can unambiguously demonstrate and prove on a scientific level—Evolution! Nobody has, as yet, collected the money and I can categorically assure you the reader no one ever will.

To call *Origin of Species* a masterpiece is to denigrate language to a disaster piece! A book that likens our ancestors to animals like shrews/mice, apes, donkeys and monkeys, is a fairy tale that does not deserve our serious consideration. And yet without any evidence whatsoever this fictitious dross has been accepted *and taught* by many otherwise intelligent people for eight generations and it shows no sign of abating. It appears to be one of the hottest topics on the internet! It is quite extraordinary that exceptional people believe this existential exotoxin to the point of accepting and believing fiction as fact.

Since I am not promoting any religion (I hesitate to write 'religious persuasion' since 99% of human beings were not 'persuaded,' they technically were forced to accept and believe their parents or guardians views!) My book could be likened to a shotgun, one barrel aimed at organised religions and the other at disorganised evolution! This is not to deny that many religions give solace, comfort, goals, fulfilment, joy, purpose and hope. The one sure thing in the past that united families, tribes, cultures and nations was a sense of particular belonging to one idealised faith. If there is a God and if he has a Messiah then the day has to come when divisive faiths, in fact, all religions shall end—be no more, and a culture of peace and love shall rule, reign and remain forever.

CONCLUSION

What I have written so far has revealed my intellectual conviction of a Creator—God or 'Cognitive Artistic Genetic Engineer.' My aversion to organised religion is irrevocable, my detestation of evolution absolute. Conviction, through incontestable knowledge, reason, facts, experiment, observation, common sense, rock solid evidence, and the provable past is the only creed that rests comfortably with my soul, mind and heart.

I could give literally seven thousand amazing chemical, biological facts from the human body and its astonishing molecular manufacturing and reproductive capabilities, but seven is enough to make thinking people pause and ponder. Before I do, I ask, you the reader, to look at each one carefully and ask seven questions regarding the biochemical marvels you are about to read.

THE ORIGIN OF SPECIOUS NONSENSE

Could this beautiful baby have come from nothing?

Is chance truly responsible for this?

Could blind unintelligent selection produce this?

How do cells know instinctively what to become?

Is a guiding intelligence in evidence?

Is design observable in the effect of cause?

Is the end result from 1 cell, not marvellous?

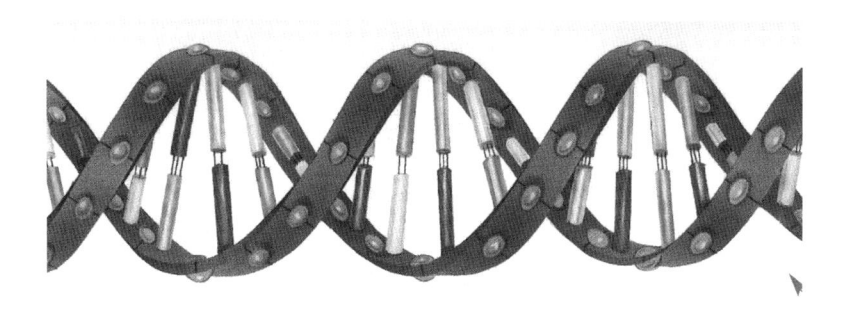

I
"Genetic programming that shames the best computer programme!"

A SINGLE CELL TO BIRTH

Almost every creature begins from a single cell. Within this one human cell, fertilised egg, are genes from both parents and the entire genetic code of each future individual. This intricate astonishing mathematically coded blueprint of over 3 billion chemical letters, 12 or so hours after conception, starts to divide and build. A human architectural project over the next 280 days (approx) is masterminded from 23 chromosomes from each parent, fused in the inner sanctum to make 46.

From one cell grow billions as each one splits, copies and carries out multiple chemical functions that silently insensibly and imperceptibly grow into human crucial organs over the next ten weeks. By that time each of us is fully formed and the size of an adult thumb. That tiny creature was once you and I receiving vital nutrients and nourishment from an excited mother as we grew in a chemically perfect environment towards our day of birth.

7 Questions

How and why do cells split?

How do toes know where to grow?

How do cells know how to build a heart?

How do cells know how to make blood?

How does blood have all the right chemicals?

How did the reproductive system develop?

Was I truly one single cell?

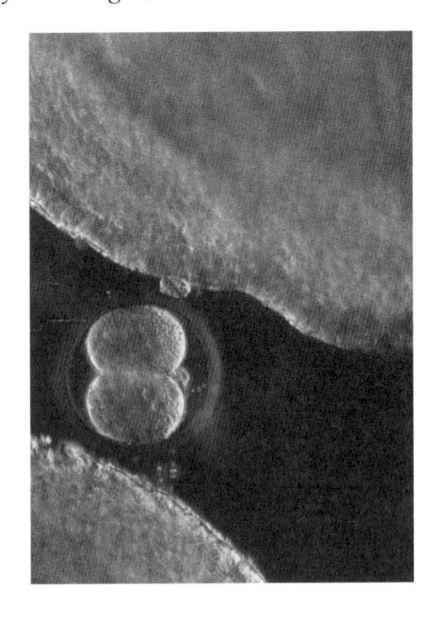

The Magical Mystery Tour Begins:
1 Cell DIVIDING into two
A Universe of Biochemical Complexity
Photo from: Lennart Nilsson's book *A Child Is Born*.

CELL DIFFERENCIATION

As cells begin to divide an astonishing thing takes place after about 7 days! CELL DIFFERENCIATION! What tells one cell it must be different from another? How do cells know how to grow? When to stop growing? What to become? Where to be placed? For example all skin cells are similar but cuticle skin is slightly different to cheek face skin! Why? How do the cells from the cheek, face, know where exactly to go? (Twice!) The molecular construction of one cell is a universe of complexity and mesmerising chemical construction. A lifetime of study under microscopic analysis and magnification would barely scratch the surface, and we are composed of not millions—not billions—but trillions: with each and every single one placed strategically, functioning economically, beneficently, silently, powerfully, producing, flesh—blood—bone- skin—teeth—eyes—tongue—toes—heart—lungs—hands—faces—brain—that enables us to soar above all species in a spectacular manner! And all this silent chemical construction has one and only one magnificent result in DNA focus! A beautiful ultimately self replicating baby.

7 Questions (2nd time)

How and why do cells split?

How do toes know where to grow?

How do cells know how to make blood?

How do cells know how to build a heart?

How does blood have all the right chemicals?

How did the reproductive system develop?

Was I truly one single cell?

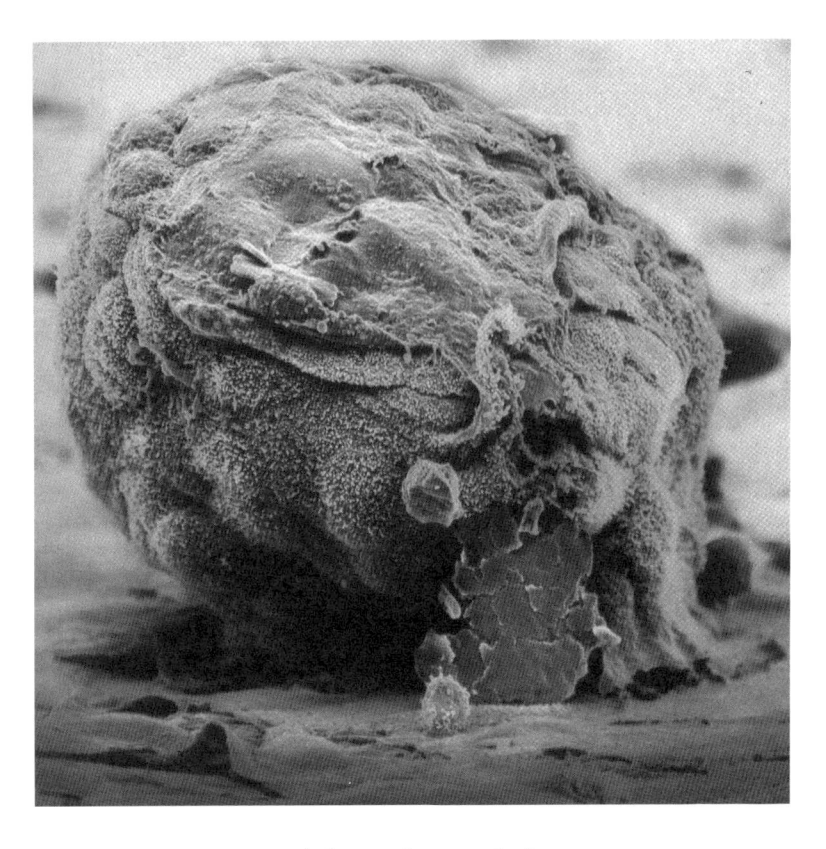

A future human being
A CELL 8 days old
A pregnant mystery!
Picture from: Lennart Nilsson's book *A Child Is Born*

EMBRYONIC GROWTH

This process alone condemns evolutionists to the mental morass of obfuscatory obscurantism.

Here is molecular order of the highest degree, design par excellence! Cells in the blastocyst four days after conception begin to differentiate; each one astoundingly knows where to go and what to grow into! For the next 10 weeks from one cell shall develop billions, all functioning brilliantly from precisely coded DNA.

Many, rapid changes must now (and do!) take place to create a favourable and sheltered environment for the growing embryo. Each of the blastocyst's signals must be complete and take place in the correct chronological order. (They do!) A careful examination of the blastosyst's surface reveals that almost every cell is unlike every other. Some have long projections, others short ones and some lack projections all together. This is called cell differentiation. Until about the eight-cell stage, all the cells look the same and serve exactly the same purpose. Now at this moment and point in embryonic time, life's true mystery happens! Its decision time and from this day on, from this blastocyst shall an embryo form, (21 days from conception!) From this shall develop a foetus! (around 63 days from conception!)

What has been happening from the true miracle of conception is staggering in its developmental highly organised march towards triumph. From one, dozens, hundreds, thousands, millions, billions, and, yes, trillions of cells all march harmoniously to the beat of two hearts in one body. They are protected in the amniotic sac, nourished through the umbilical cord, feeding from the blood rich placenta. Waste products are removed in a hygienic and highly organised fashion to the healthy benefit of mother and baby.

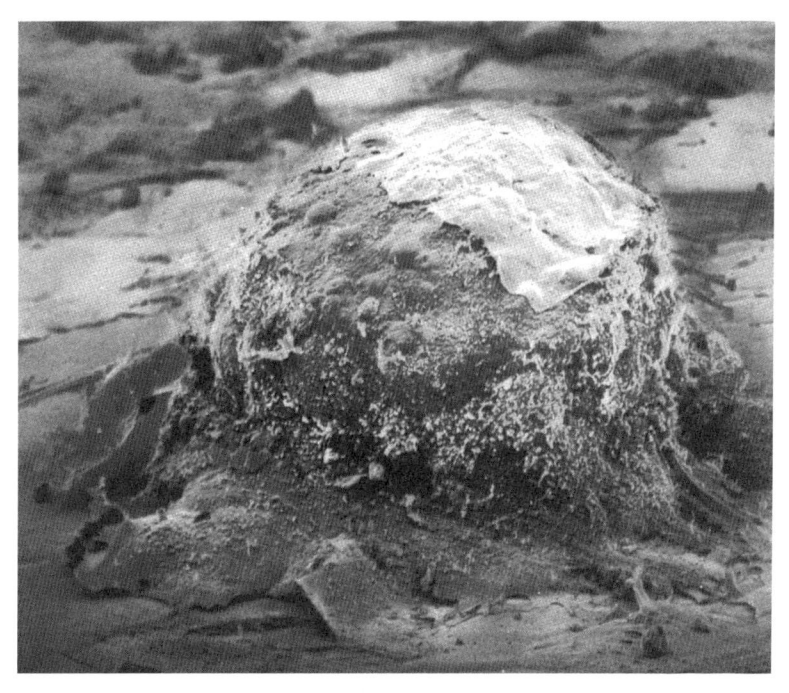

A future baby. . . .
A past miracle . . .
BLASTOCYST
Picture taken from: Lennart Nilsson's book, *A Child Is Born*

It is sacrificing reason on the altar of treason to accept that the greatest construction of all time—a human being with a brain is the result of chance, random selection and destructive mutations. It is the irrational 3-legged chair of hopeless speculation that bears no resemblance whatsoever to reality and observable functioning perfect order.

Millions of molecular marvellous mysteries are chronologically taking place every 24 hours and inexorably leading to a welcoming exit after about 280 days:

The triumph of one cell metamorphosing into one beautiful baby. We are free to believe what we want, but I have chosen reason.

7 Questions (3rd Time)

How and why do cells split?

How do toes know where to grow?

How do cells know how to build a heart?

How do cells know how to make blood?

How does blood have all the right chemicals?

How did the reproductive system develop?

Was I truly at one time one single cell?

8 WEEKS

The time schedule of physical development is precisely programmed and varies little between individuals, although the genetic blueprint may vary considerably in other respects.

At eight weeks the embryo is about one and a half inches long and every single organ is in place. Everything found in an adult is now in the foetus as it is weightless and suspended in the amniotic sac for protection.

One hundred thousand (yes 100,000) nerve cells are miraculously manufactured every minute (yes, every minute) and by the time the baby is born, there will be 100 billion (yes, 100 billion) in the brain.

How is this wondrous creation happening? Scientists who are lucky enough to observe all this can still only describe the process and speculate as to causes.

For example the heart is now fully formed and beating twice as fast as its mother's and is driving the foetal blood to and from the placenta. This astonishing organ can beat and pump 75,000 pints of blood a day, through 100,000 kilometres of blood vessels for 100 years without skipping a beat. If 100 of the worlds best inventors/engineers spent 50 years trying they could not create a muscular pump the size of a fist that could push that volume of liquid without stopping for 70–80–90 and sometimes over 100 years.

At eight weeks not only the heart but every single other organ is formed, functioning and self testing. When each organ is microscopically examined and observed, it is patently obvious extraordinarily superior unmatchable intelligence is responsible and there is no other logical reasonable rational realistic reliable explanation.

None . . .

At around seven weeks a future person who is not even seven ounces will at time of exit be about seven pounds and grow to seven stone in seven years, in some cases. It is difficult to imagine that you and I were at around two months the size of a thumb. All the body's amazing systems are now interconnected and growing stronger through cell multiplication. Time, protection, nourishment and cautious care are all that are needed to bring this future parturition to triumphant fruition.

It is irrational to suggest that such molecular motion, chemical cleverness, D.N.A. codes, sperm and egg, 46 chromosomes, cellular differentiation, hormones and blood, skin and bone, eyes and heart plus millions of other atomic structures came from nothing—means nothing, will be nothing! And since it is totally irrational I commit it to the realm of ridiculous speculative fantasy. It is far more reasonable to conclude a creator of awesome prodigious intellectual capabilities was—is and forever shall be . . .

The Cognitive Artistic Genetic Engineer. (God)

7 Questions (4th Time)

How and why do cells split?

How do toes know where to grow?

How do cells know how to build a heart?

How do cells know how to make blood?

How does blood have all the right chemicals?

How did the reproductive system develop?

Was I truly at one time one single cell?

THE PLACENTA

No placenta, no baby! On the first cellular division, one is for baby and other for the crucial placenta. From an indistinguishable cluster of cells grows this vital connected rainbow of nutritious raindrops. Temporary organs in the mother's body are responsible for the entire interchange of nutrients to and waste products from the embryo. Also most bacteria and viruses cannot pass through here to harm the foetus because of a strategically positioned placental filter/barrier. From this life enhancing chemical mechanism the embryo's blood absorbs proteins, fat and sugar for the constant work of cell building, as well as oxygen to fuel the process.

The growing life is oblivious to the marvels occurring and the mother pleasantly aware of her expanding tummy. It is time to pause, reflect and reason. How is it possible that a completely separate—similar human being can come from a man and woman? Before pregnancy no placenta, during pregnancy no life without placenta and immediately after birth the placenta is expelled and of absolutely no further use. (Stem cells is a peripheral issue!)

In simple language the blood rich placenta is absolutely crucial to the healthy growth of the foetus. It can be likened to an opening parachute strapped to a man—who has jumped from a plane; it means his life. And as at today's date, 2010, 7 billion chemical parachutes opened, facilitated life and were appreciatively discarded.

7 Questions (5th Time)

How and why do cells split?

How do toes know where to grow?

How do cells know how to build a heart?

How do cells know how to make blood?

How does blood have all the right chemicals?

How did the reproductive system develop?

Was I truly at one time one single cell?

JOHN J MAY

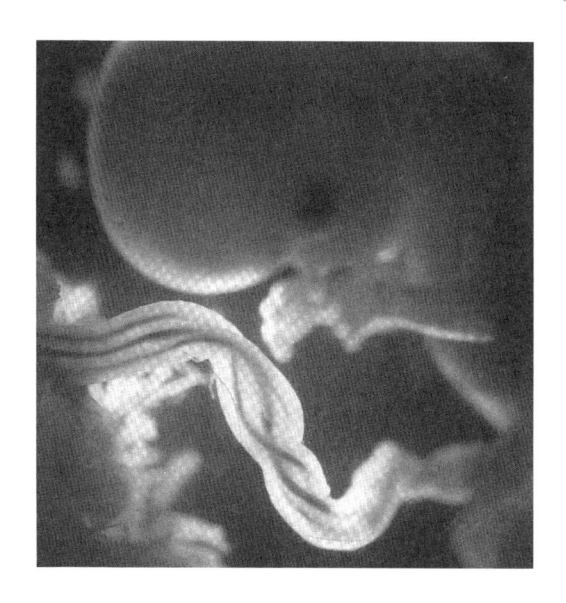

CODES DEMYSTIFYING CODES
Picture of the placenta attached to the baby.
Picture taken from: Lennart Nilsson's book, *A Child Is Born*.

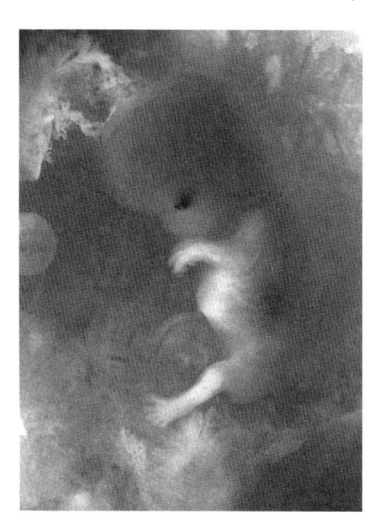

OUR FIRST HOME
Picture of a baby at 3 months.
From *A Child Is Born*.

THE ORIGIN OF SPECIOUS NONSENSE

This three month old foetus has everything it needs. It is a space traveller in a tiny capsule complete with lifeline. The outer ragged halo is the thin tough chorioid sac, enclosing the amniotic fluid where the temperature is 99.5 degrees Fahrenheit, a little higher than mother, but perfect for baby.

The vital yolk sac has now served its purpose as the blood cell factory, and now the liver, spleen and bone marrow seamlessly take over mass production. The foetus now lives pleasantly in its ideal environment the growing womb, and is virtually germ free and further protected by a chemical plug at the neck of the uterus which fights off potential infections.

Creation proceeds silently and insensibly and the face is now under construction even though it is only about 20 grams (one ounce)

A chemical artistic individual process is relentlessly, cognitively, beautifully building a new unique human being.

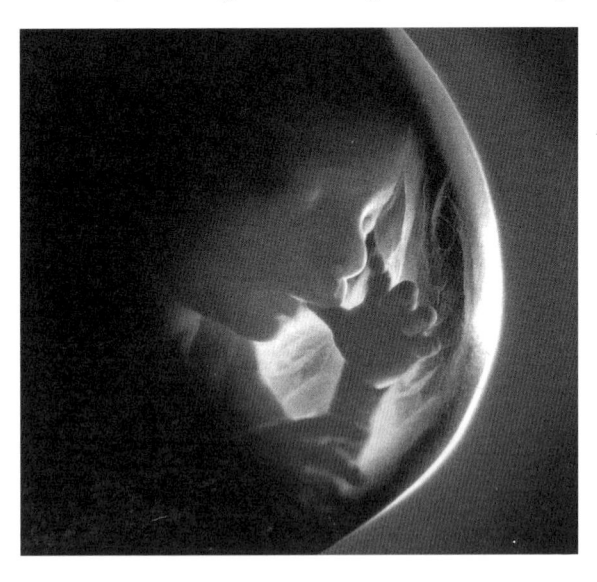

From a single cell
Lennart Nilsson's book, *A Child Is Born*.

The father is oblivious to developments and mother is hardly conscious physically of growing life embedded in her flesh and anchored to her heart. Here is love's reward, life's prize, young dreams fulfilled, hope's home—now the mystery of life unites two people producing three.

A medical, chemical, physiological or photographic textbook can never capture the hourly, daily, weekly, monthly march to its final goal—birth. There is no human feeling to match the joy we experience at reproducing ourselves, none.

7 Questions (6th Time)

How and why do cells split?

How do toes know where to grow?

How do cells know how to build a heart?

How do cells know how to make blood?

How does blood have all the right chemicals?

How did the reproductive system develop?

Was I truly at one time one single cell?

30 FETAL DEVELOPMENT NEEDS

Oxygenated blood.

White blood cells to fight infection.

Placental hormones.

Copious amounts of nutrients.

Bone marrow.

Operational placental filter.

Chorioid protective sac.

Yolk Sac—blood cell factory (for first 11 weeks)

Amniotic fluid.

Slightly higher temperature.

Liver—spleen and bone marrow, blood cell production takes over from yolk sac after 11 weeks.

Neck of cervix blocked to prevent infections

To form the face in 4th month, five "processes" are absolutely crucial, and are one by one correctly placed.

Long nerve paths extend from each eye and must cross before reaching the visual cortex of the brain.

The eyes and brain play chemical ping-pong to produce 20/20 vision.

The ear is formed from 3 components developing from three different directions during the construction process.

The hand forms before the foot.

Arms and legs lengthen.

The umbilical cord grows an ingenious safety device which stops it kinking and so prevents blood supply being cut off.

Cartilage waits to become bone . . . later.

Protective skin ointment protects baby.

Male and female sex organs develop precisely as designed and for a noble purpose . . .

Precursors of sperm develop.

By the 5th month five hundred thousand eggs develop and are surrounded by nutrient cells.

Chemical "policemen" prevent the mother from ovulating during pregnancy (and her female baby for the next 12/13 years.)

At 6 months the hand is a tiny work of art, and the thumb sometimes touches the lips and simulates sucking.

At 7 months a protective layer of fat builds up on the dermis.

Hundreds of billions of brain nerve cells formed and functioning.

THE ORIGIN OF SPECIOUS NONSENSE

Quantity of amniotic fluid increases as needed towards end of pregnancy.

The foetus can swallow this fluid and gives the alimentary canal some vital practice.

In liquid/water but not drowning!
From Lennart Nilsson's book, *A Child Is Born*.

I have noted only thirty but there are hundreds of molecular orderly precise functions on a cellular level taking place every 60 seconds. As a simple illustration imagine counting one thousand numbered steps up a small mountain. Each one is strategically placed as the sides of each are like a cog and can only lock into the one below, the one above and nowhere else. Eventually the 1,000 are precisely built and a sturdy safe walkway has been constructed. Now look at your thumb and realise that for the cells to, first of all be in place, to build the bone, skin, nail, cuticle, veins, knuckle, fingerprint, shape, ligaments, weight, colour, connective tissue, nerves, et cetera, tens of thousands of steps needed to be taken and all in order, in place and in time. (And that's just our thumbs!)

The point being: None of the above happens by chance. Ignore God if you wish, curse him if you want, but stop pretending pregnancy promotes evolutionary concepts when reason and evidence clearly show the opposite. Mental dysfunction manifests itself clearly through disassociation from reality and evinces shades of psychosis. I think the epithet most descriptive of intelligent individuals who embrace evolution and reject reason is FANTASISTS.

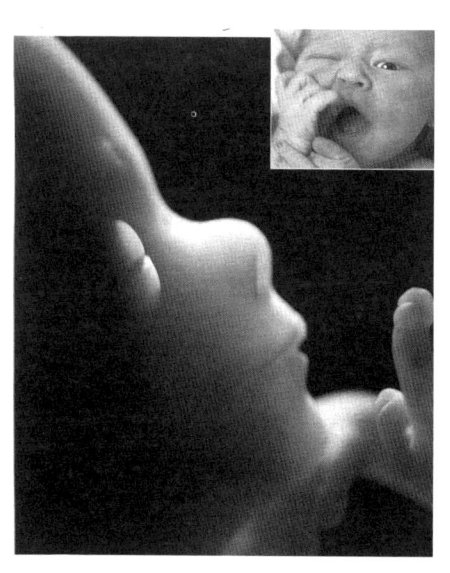

Peace in the pumping heart and restful growing womb.
From Lennart Nilsson's book, *A Child Is Born*.

LOVING GROWTH

At two months the design work is complete, all its organs have been formed. From now on the embryo must grow, develop what has been created, refine its functions and test its systems. It now graduates to a new stage and becomes a foetus.

The foetus now weighs roughly 13g (less that ½ an ounce). In 50 days it has been transformed from a single cell into many millions, **all precisely programmed for their specific tasks.** It is at this moment that I ask you, the reader, to stop reading and visualise something the size of a pinhead. At its core is the genetic blueprint that starts to organise genes, cells, DNA, cell differentiation; in fact everything on a molecular level to grow a little perfect baby! How this fantastic 50 day development is governed is still, in many respects, a mystery. In every cell of the eye, hand and fingers, there are about 100,000 genes. *How* does the cell *know* that it is to become part of the cornea, the lens, the vitreous body, the retina or optical nerve? *How* does it *know* which genes are to be *used* at each crucial moment, and which are to be *excluded?*

The mother/woman bearing a tiny budding life, wondering at everything that is morphing in her body need not concern herself with these questions. Development follows a genomic pattern laid down thousands of years ago through a biochemical engineer of prodigious and unsurpassable intellect, and her contribution is to live in such a way that this pattern of molecular growth is not disrupted.

One thing we can be absolutely certain of, evolution is not, was not and cannot be responsible for such awesome step by step, time clock orderly building block of unerring precision. How could it? It is supposed to have come from nothing and built up our cosmos including every single living thing! How did it accomplish all this precision, beauty and order?

7 Specious Answers

Accidents (chemical)

Chance (molecular)

Mutations (dangerously damaging)

Blind selection (un-natural)

Spontaneous Generation (never happens)

Impossible odds (mathematically)

Random accidents (produce disorder inevitably)

The body and brain, in particular, are chemical creations of such astonishing intelligent brilliance that reason by its very nature convinces us of a First Cause we call GOD. Our sensible acknowledgement of His obvious existence is not contingent on our flawed perception of his unique moral character, fascinating personality, or puzzling purpose! Now the foetus gains strength through nourishment and the excited woman already loves her baby. Some 95% of all babies are born between the 266th and 294th day and there is not one of those that the woman is not thinking so many questions about her miracle growth. The English language cannot adequately describe the joy she feels at her baby's first tingly kicks, except to say she is ecstatic, and completely filled with inexpressible pleasure.

BIRTH

"Birth is difficult and dying is mean, so let's get some lovin' in between"

The above words are from an old blues song but it captures perfectly life's struggle for meaning and more importantly, love.

The sadness, shortness and bleakness, if we are lucky is occasionally interspersed by magic moments. And none are more mystical, joyous, life enhancing, or thrilling than the long awaited arrival of a baby. None. None.

About a month before delivery to make the birth easier the foetus rotates for a head first departure in 97% of cases. In the modern world most women can look forward to giving birth as a powerful and exhilarating experience matched by no other in life.

Labour is the time every daughter asks her mother about! It is basically in three stages. Dilation—Expulsion—Placenta purpose over.

"Being born entails *considerable stress for the baby*. During each contraction, when the placenta and umbilical cord are compressed as the uterine muscles draw together, the supply of oxygen to the baby is curtailed. Initially, the baby's pulse slows down during intervals. *The baby has a tremendous capacity for withstanding strains,* and here the *adrenal glands* play an important part. These glands secrete large quantities of *adrenaline* and *noradrenaline—hormones that are important in protecting the foetus in the event of an oxygen deficiency,* since they promote the heart's pumping capacity, speed up the heart rate, channel blood to the sensitive brain and raise the blood sugar level. *Never again* in later life are such large amounts of these stress hormones secreted, and this indicates *how stressful it is to be born,* but also *how well prepared the baby is for this stress*. The hormones are also important in preparing the lungs for life outside the uterus. In particular, *adrenaline reduces the formation of liquid in the lungs* that has taken place throughout the life of the foetus, and *expands the respiratory tract.*" Also, "Now pain, joy, excitement are mingled. The experience of pain varies from one woman to another, depending on sensitivity, expectations and anaesthetic effects. *For the baby, birth itself is a cataclysmic event. The adrenaline shock* counteracts the *oxygen deficiency* in the final stage, and prepares the baby for *the sudden switch to breathing through the lungs.*" From Lennart Nilsson's brilliant book, *A Child is Born*.

If the father is present during labour this enhances a woman's sense of security. Her body also produces its own pain relieving hormones—endorphins which help dull the perception and reality of pain.

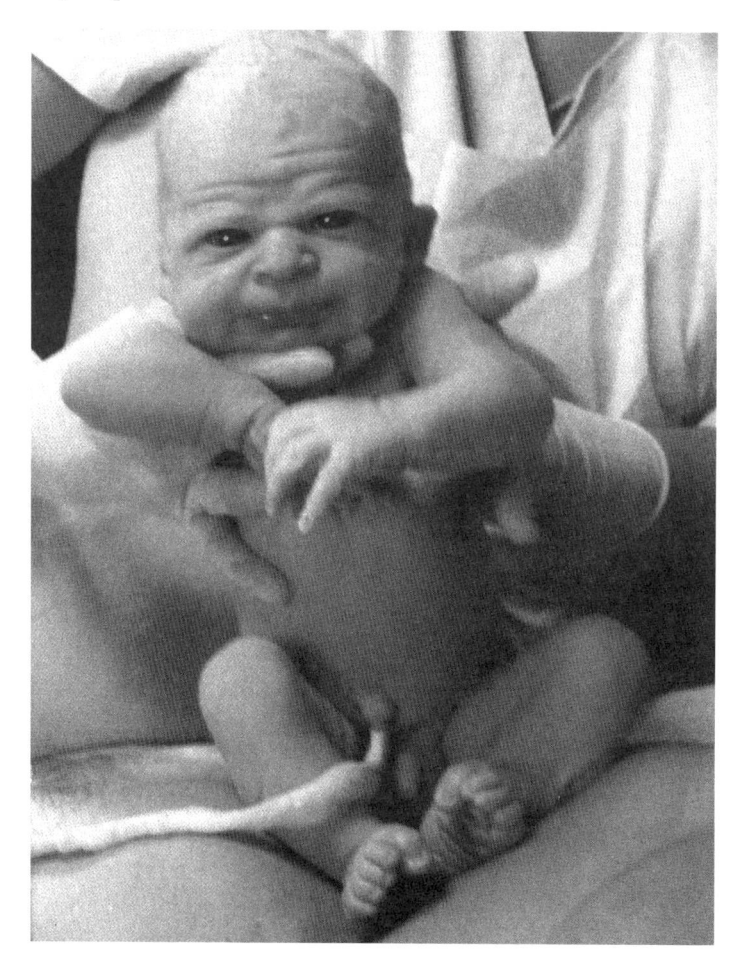

I'm ALIVE and I've Arrived. . . . .
(From one cell)
Picture of a baby moments after birth.
From Lennart Nilsson's book, *A Child Is Born*.

THE DAY OF BIRTH

Curiosity, longing, wonder, questions, anxiety and delight are some of the many thoughts feelings and emotions coursing through her brain, flesh and blood. About 280 days of dreamy dawning dazzling joyous dopamine. In 97% of all pregnancies, around four weeks before delivery the "baby" rotates spontaneously and from top to bottom, positions its head down the birth canal to arrive, alive, as this makes delivery easiest. Question—how do almost 100% of babies know when and how to turn and place their malleable heads into exactly the correct position for an easier and safer delivery?

The future mother is now finally preparing for a powerful and exhilarating experience matched by no other in life. Labour pains arrive and labour is experienced by awkwardness, tension, trepidation, tiredness, alarm and hope. Soon the mother will deliver her very own tiny baby which will receive nourishment/milk from her warm tender breasts;

Astonishingly the 7 most necessary food items for newborns are:

Antibodies—To fight infections. (Present)

Minerals—For strong bones and cellular growth. (Present)

Lactoferrin—Prevents the growth of iron dependent bacteria in the gastrointestinal tract which inhibits certain organisms, such as coliforms and yeast, that require iron. (Present)

Secretory IgA—Works to protect the infant from viruses and bacteria, specifically those that the baby, mom, and family are exposed to. It also helps to protect against E. Coli and possibly allergies. (Present)

Lysozyme—Is an enzyme that protects the infant against E. Coli and Salmonella. It also promotes the growth of healthy intestinal flora and has anti-inflammatory functions. (Present)

Bifidusfactor—Supports the growth of lactobacillus, a beneficial bacteria that protects the baby against harmful bacteria by creating an acidic environment where they cannot survive. (Present)

Fats, Proteins, Vitamins, Carbohydrates—All found in mothers nutritious milk and approximately 60–80% of all proteins in human milk is whey protein; these proteins have great infection protection properties. (Present)

The curtain is about to go up on life's greatest drama, the props are in place, the stage is set. We now are about to take our first breath! (or die!) And when it does, it has 100 billion nerve cells in its little brain. "At last, out in the real world—a world of dazzling light, cold air and loud noises. Almost immediately comes the first cry—a rare, sometimes hesitating sound. Now 25 million little air sacs (alveoli) must be filled with air. Up to now, they have held fluid, but this is rapidly expelled in blood and lymph. The first breaths are among the most arduous of one's whole life. The bloodstream must now be redirected. *The hole in the partition between the heart's atria is **SEALED**.* (*For that hole to seal at that moment* to immediately facilitate the vital blood flow in the correct direction through the lungs—just at *that time*—is a miracle of miracles).

Up to now, oxygen has come from the mother via the umbilical cord, but now the baby is self-sufficient: the blood must be directed into the lungs and then all over the body. For the parents, this is a miraculous moment. Such a tiny human being, so full of life—their very own child!"

Euphoria—exhaustion—joy—tears—rapture—amazement—wonder—triumph—tenderness overwhelm the mother and transfix the father. Here is an island of joy amidst the bright, noisy, sterile hospital environment.

THE ORIGIN OF SPECIOUS NONSENSE

THE MAGNIFICENT RESULT OF MYSTERIOUS LOVE.

It is quite simply not credible that this beautiful baby combining the physical characteristics of both parents, plus linking and strengthening two humans into three in love came from nothing!

CONCLUSION

Once again I ask you the reader to reflect on the fact that the baby in the photo, all babies, you and me, once were smaller than the point of the pen I now write with. Our lives are busy and meditating on this may be something we are not used to doing! Be that as it may, I have presented my case for reason against "pie in the sky." Humans are not cognitively aware of the law of unintended consequences and have a fatal capacity for allowing ourselves to be deluded by glib mediocrities. There are dozens of anthropological fantasies to choose from, some unfortunately have imperceptibly become part of our psyche through cultural percolation. Believing we are only a higher form of animal condemns us to a lower form of living. This is inevitable since it diminishes the seven fundamental characteristics of what it means to be human.

LOVE

HOPE

MORALITY

KINDNESS

JUSTICE

PHILOSOPHY

SPIRITUALITY

Of all known theories, evolution is possibly the most destructive, since it claims life arose through chance in the ancient past. We know now from science this is an utter impossibility. The present means nothing more than meandering madness since in the view of scientists it sprang from ancient mud! (I kid you not!) Saddest of all, they deliberately spurn hope for themselves and their children's future! This is the seed of destructive depression in millions of agnostics and atheists. Their supercilious disdain of 'innocent/ignorant' religionists who harbour harbingers for a future joyous life is quite shocking. They are like turkeys voting for Christmas.

Nothing grows without a seed and nothing develops without a stalk. In the womb this is the vital temporary miraculous bridge between baby and mother, the miraculous placenta. An amazing fact—from this lifeline, embryo's blood absorbs vital protein, fat and sugar which are all necessary for cell building plus oxygen to feed the process. If cells don't build, life cannot grow and for people who are sufficiently curious I insert the following two pages from *Atlas of the Human Body* by Vigue-Martin published in 2008 by Chartwell books Inc. USA.

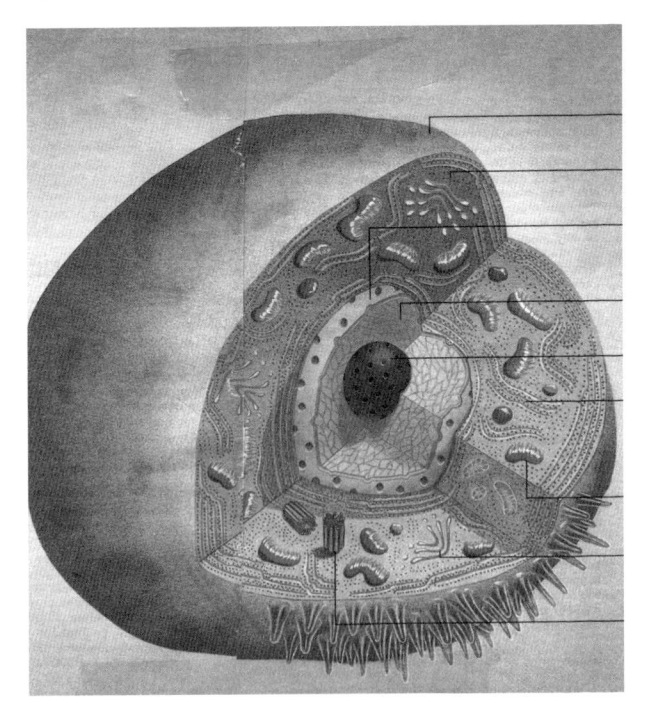

The Mysterious Magical Cell. Watchtower Society U S A

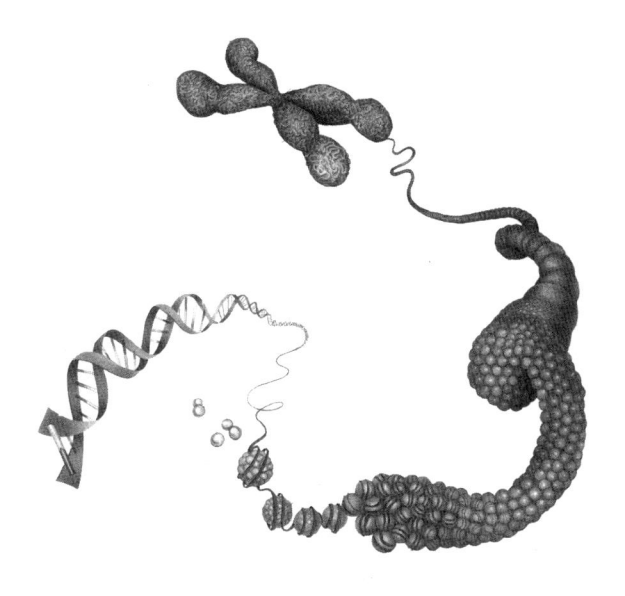

This chemical code was once you and I . . . Vigue-Martin.

The ontological argument is facile since it is quite evidently beyond our mental capabilities to conquer and grasp even one cell, which would need 42,700 x 600 page books of small print to try explain, so how can we hope to ever comprehend the mind of THE GREAT SCIENTIST (God). Evolution should never ever be taught in a biology class since it is a metaphysical concept with its roots in fantasy. It possibly is more suited to philosophy since formerly alchemy, astrology and astronomy were studied as natural phenomena and were systemised in theory and experiment. Nevertheless the bottom line in this ludicrous debate that debases its deluded deceived disciples and poisons with pernicious consequences all who embrace it is—swallow this concoction at your peril.

Every time I hear the laughable oxymoron, "Evolutionary Biology" I harness it to its twin; "Santa Claus Biology" because the way it is linguistically meant is not the way it scientifically happens. As I look around my home at the many interesting and beautiful objects, (the vast majority are not expensive) I know for an absolute fact, that each and every single object, art and mundane had a maker, artist, builder or designer. The argument from design has never, I repeat, never ever, been dismantled. To counter the argument of effect from cause with 'who made the cause?' is a smokescreen for, no argument at all.' Factualists are consistent where-as, fantasists are contentious conceptual con-artists.

I have written many words to present my case and if 0.1% is erroneous, I present a quote from Stephen Jay Gould, American author, biologist, science historian, atheist. (1941–2002) "Error is the inevitable by-product of daring." I have dared to expose thousands of scientists, biologists, anthropologists, zoologists, and many more besides, who have childishly bought into a fiction and established that 99.99% of what they say re: evolution is in scientific fact, **ERROR!**

Very few humans enjoy the experience of being corrected. One reason being pride! Do I have it? Of course I do. However, I have enough love of truth to openly admit, I am an ignorant man and the more self educated I become the more ignorant I realise I truly am. Nevertheless, I will listen to anyone who presents factual evidence on any topic and if my views are wrong I shall instantly correct them. I have no loyalty to childhood teachings, traditions, customs, views or comforting fictions if they might be harmful.

I view life as a treasure, family as my purpose, friends as a gift, the future uncertain and my chosen God my joy. It is a thrilling experience to have been born on this earth, a travelling open-air spaceship. In fact while we sleep for eight hours each night our "spaceship" has rotated east to west 8,000 miles and silently speeding round the sun at 66,000 miles per hour. We have also travelled while we slept—858,000 kilometres—silently, quietly, fantastically and it rarely loses one second or goes out of orbit by one millimetre! Oh, and it needs no rocket fuel! I have been extremely lucky as a gentile to have discovered through the Jews that the real purpose God (Yahweh) originally formed man and woman and placed them on this amazing earth ...

OUR ETERNAL HOME

As I wrote earlier, I am implacably opposed to organised religion and irrevocably at war with evolution. An unusual position to take some might say! Not really, I respond: Organising people to love God (or gods!) strangles spontaneity and inevitably produces artificiality, superficiality and non-spirituality. My hatred for evolution is as deep as my detestation of war and those who cause it! (Blessed by various clergy). My reasons are simple. It is a toxic virus that paralyses reason, chokes common sense, rejects facts, denies logic, elevates confusion, spurns evidence, grasps at straws while ignoring flaws and suffers delusion's that illusion's are reality.

There are none so blind as those who will not see. However, the tragedy here is they do see, but what they see is invisible to their reason. Hence the vocal quasi arguments and papers interminably promoting the impossible: evolution.

D-N-A

DNA is the Achilles heel of evolution. It is a specialist language that laughs at those who are lucky enough to examine it in some detail and foolish enough to proclaim, "This highly organised tightly bound mysterious molecular marvel methodically arranged itself." Bill Gates, co-founder of Microsoft, with Paul Allen in Albuquerque New Mexico 1975. said, "DNA is like a software programme, only more complex than anything we've ever devised." *Anything*, he said. So if a software programme takes incisive intelligence to design (all you software engineers out there know exactly what I am saying) why is it not reasonable to conclude that DNA—THE masterly genetic code, also proclaims a Cognitive Genetic Artistic Engineer? Are certain complex computer codes not intelligent?

Evolution tells us that through chance, mutations and natural selection, living things evolved. Yet to evolve means to gradually change certain aspects of some living thing until it becomes another type of creature, and this can only be done by changing the genetic information. And THAT is the insurmountable problem of irreducible complexity. It cannot be overturned or breached, dismissed, or ignored, dislodged or disproved, and all the specious disingenuous disinformation is disgraceful the in face of facts.

Once the penny drops, the truth is, there is no evidence to support the heart of evolution which is, the airy-fairy fantasy of one species slowly evolving into a different species. When that knowledge seeps into one's mind, the heart rejoices at the only other possibility, that a creator of prodigious intellect, astonishing artistry, genetic genius and engineering brilliance is the God of the cosmos. Now the question needs to be asked—how are so many taken in by so little? The answer is confusion in religion and the certainty of ignorance. Most people are simply too busy to give it much thought, and whenever the topic comes up they shut up to whoever shouts up. Richard Dawkins (whose intellect I admire) wrote in his book, *The Selfish Gene*, in his chapter titled: 'The Long Reach of the Gene,' page 261; "Living things, of course, were never designed on drawing boards.

But they do go back to fresh beginnings. They make a clean start in every generation. Every new organism begins as a single cell and grows anew. It inherits the *ideas* of ancestral design, in the form of the DNA program, but it does not inherit the physical organs of its ancestors. It does not inherit its parent's heart and remould it into a new (and possibly improved) heart. It starts from scratch, as a single cell, and grows a new heart, using the same design program as its parent's heart, to which improvements may be added. You see the conclusion I am leading up to. One important thing about a 'bottlenecked' life cycle is that it makes possible the equivalent of going back to the drawing board."

Three times he mentions 'Design' and twice the fact that every new organism begins as a single cell. Once, The DNA programme and three times in one paragraph of 138 words, he adopts Charles Darwin's fantasy language.

"Possibly improved"

"May be added"

"Makes possible"

A careful analysis of this paragraph reveals; firstly he has absolutely no idea of how flesh—blood and vital functioning reproductive systems arrived on this planet! Speculation is not evidence. Supposition not proof.

He next glibly mentions; 'Fresh beginnings' as if 'they' were ever stale! He is right by default when he states, 'They' make a clean start in every generation. Then knowledgeably writes, "Every new organism begins as a single cell and grows anew." One needs to stop here and realise the import of what he wrote . . . every single organism, et cetera. Go now, to any medical textbook and spend some time examining *just one*, the *human organism* and remember what he said; this staggering organism grew from one cell—and try grasp the fact that no biologist or scientist on this earth truly understands or can fully explain the depth, mystery—complexity and codified molecular certainty in just this one magnificent cell which morphs into our:

SKIN

BLOOD

BONE

HEART

HAIR

TEETH

BRAIN

AND THEN

The Muscular system

The Skeletal system

The Cardiovascular system

The Digestive system

The Respiratory system

The Urinary system

The Reproductive system

The Glandular system

The Nervous system

The Sensory system

Each organ and system depends on all the others to function correctly and contribute to the physiological balance that makes us human beings. To repeat what he correctly wrote; "Every new organism begins as a single cell and grows anew."

It's easy to write, easy to read and easy to forget! However it becomes unforgettable when we concentrate the mind on this dot " . ," meditate that in the past we were smaller than this dot " . ," and that we anatomically grew imperceptibly in our mothers bodies from a chemical dot, to a lovely living gift for the mother, father, delighted grandparents, brothers and sisters. It many times, looks walks and talks like its parents and all from one tiny astonishing dot " . ," (fertilised egg-cell.)

It is not possible nor reasonable to conclude that this baby from this tiny dot " . ," to this adult to reproduction once again from a miniscule dot " . ," designed itself! It is simply ***not credible.***

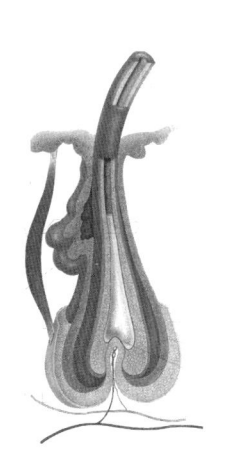

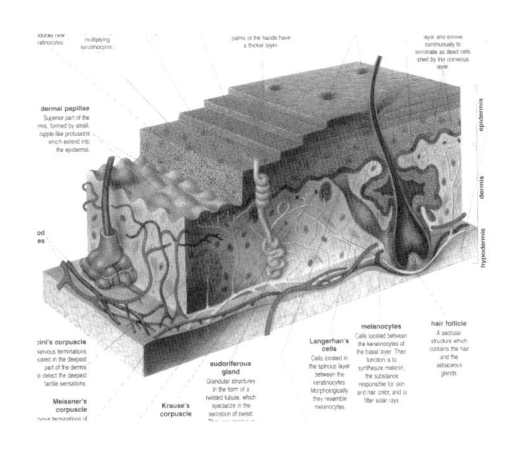

Human cells
Hair follicle and skin
Duplicating and copying us
Hair for beauty and skin for life!

THE ORIGIN OF SPECIOUS NONSENSE

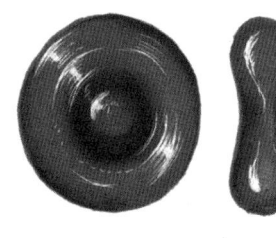

Human red blood cells
12 Pints of molecular liquid life

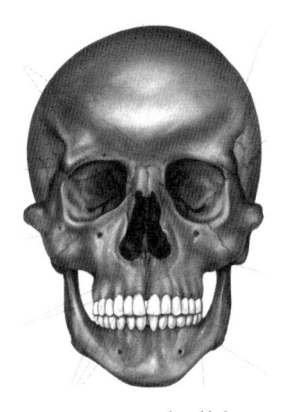

Human skull bones
Unique and obvious

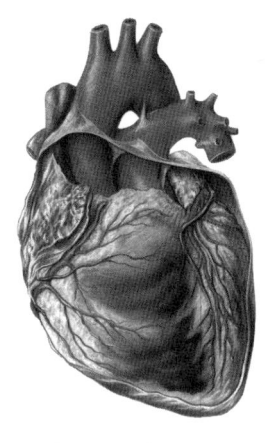

A human heart

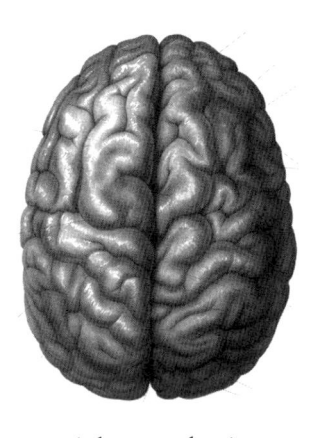

A human brain

It stops we die. It can beat for over 100 years both day and night!

The mechanism by which we Love, Live, Laugh and hopefully REASON!

Every single species of animal we look at, every fish or bird all started from a chemical certified code smaller than this ". " dot. . . . Every banana, orange, pear, apple, pineapple, melon, nuts, figs, pomegranates with their many pips and delicious fruit all came from tiny seeds. This means in the everyday language of computer programmers; 'A high-level language in intelligent code is crucial to every single species of life and food on this planet.' In reality the scientific consequences of this molecular fact is—every egg—sperm—seed and cell can ever only become what the code/language 'speaks' and nothing else. They are the facts and that is the truth. A code can be broken, examined, copied, discarded, ignored, admired, even worshiped! But the one thing a code cannot do if it is programmed to become, from specific genetic information, say an amino acid, is become a heart! Amino acids are essential components of proteins* and vital for healthy growth, but alone can never ever grow into a human heart . . . and why?

Because of individual specific genetic codes.

Richard then discusses inherited design as correctly connected to the language of the specific DNA programme and obviously notes a species does not inherit the physical organs of its ancestors. He then tell us (I kid you not!) "It inherits the *'ideas'* of ancestral design." In June 2000, the hereditary code of life, the Human Genome consisting of all the DNA of our species was publicly revealed to the world by scientist, Francis S. Collins, (a Christian) and his many associates in the USA and around the world. This extraordinary four letter text was over 3 thousand million chemically connected letters long. Not one, I repeat, not even one was connected to what Richard calls "ideas!" Ideas come from the mind, organs from codes. Yes, he does attempt to qualify his statement by saying "It inherits———————ancestral design in the form of the DNA program." I have deliberately left out his three superfluous and unscientific words "the ideas of" to highlight unscientific inaccuracy. One likely reason he chose the descriptive term "idea" is because of Charles Darwin's so called 'Big Idea' which is the easy language of evolutionists.

However, thoughts, conceptions, ideas are not chemically in the DNA code.

In the middle of the paragraph he amazingly suggests; "It does not inherit its parent's heart and 'remould' it into a new (and possibly improved) heart." It is only when one stops and thinks about this that the insanity of it strikes.

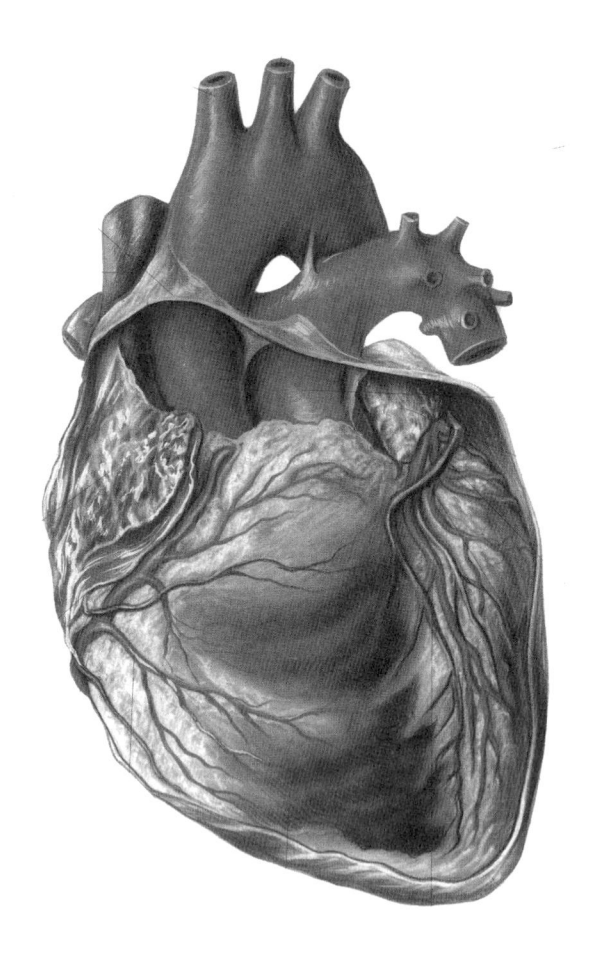

THE MOST POWERFUL PUMP ON EARTH
HUMAN HEART

Richard is trying to tell us that a growing heart in the womb can from its chemical pattern (mould) stop moulding and start re-moulding to a different pattern or design and "possibly improve" on the existing code as it develops a perfect functioning heart with its multitude of interconnected powerful muscular components.

We are not talking about car tyres here. No, we are discussing the most powerful pump ever created-designed in history. He then compounds this unrealistic scenario with further fantastical fanciful notions, "It starts from scratch, (correct) as a single cell, (correct) and grows a new heart, (correct) using the same design programme as its parent's heart, (correct) to which improvements may be added." FALSE. Improvements! What improvements? To which species? Will someone please tell Richard that the hearts of humans, lions, elephants, whales, giraffes, eagles are all perfectly functioning. Thank you.

He then asks his apparently credulous readers, "You see the conclusion I am leading up to!" Well, yes, I do, and it is quite simply laughably ludicrous. His concluding sentence states, "One important thing about a 'Bottlenecked' life cycle is that it makes possible the equivalent of going back to the drawing board." The word bottleneck basically means obstruction. Biologists are unaware of any so called 'obstruction' in the molecular mechanism of a methodical calculated chemically developing species in its embryonic 'life cycle.' In simple language a fertilized human egg always becomes in predetermined time, a baby. He then suggests that this unbelievable fantasy of 'Bottlenecked life cycles' "goes back to the drawing board." Has he forgotten what he opened his paragraph with, "living things, of course, were never designed on drawing boards."

If "living things were never designed on drawing boards," then why does he blatantly contradict himself at the end of this paragraph by suggesting the 'importance' of "Bottlenecked life cycles . . . going back to the place he stated NEVER happened . . ." design on drawing boards?

No, Richard, you cannot have it both ways. Your speculations and contradictions are everywhere in your erudite writings. The reason you have gathered so many disciples is most people think you know what you are talking about. And I have only analysed one paragraph out of hundreds. Confidence is not certitude. Erudition not facts. Speculation not proof. Faith not evidence. Ideas may not be reality. And finally Darwinian evolution remains reason's revolution

On July 1st, 2009, I attended in the Science Gallery Lecture Hall, Trinity College, Dublin along with my brother Gerard, a meeting with the theme—"Is Religion an Infection?" The panellist's were Dr. William Reville, a Roman Catholic and genetics professor, and Dr. David McConnell, athiest and biochemist professor.

Other notables on the panel intelligently contributed to the discussion. Naturally the idea of evolution reared its ugly head and I spoke up and clashed with a few people present. The following is astonishing out of an audience of 150. I made the factual claim that not one, I repeat, not one scientific fact can be presented to unambiguously substantiate the claim that; "one species slowly evolved into a different species!" (Which is the heart of Darwinian evolution!) There was a roar of disapproval from about half or more in the audience. I was in the front row, stood up, turned round and faced the audience, then in a loud voice said; "Any person here tonight who can present me with one scientific fact, to prove that one species ever changed by incremental mutations into a different species, I shall reward with €1,000." At the interval—no one came up to me! At the conclusion—no one walked up to me! During the meeting—no one successfully contradicted me!

Not one out of *at least* 150 people approached me! The reason is simple. People are not used to hearing an enemy of organised religion denounce vehemently—logically and scientifically the hoax—myth—virus and poison of evolution.

This scenario could be played out in any lecture hall of any university worldwide. The time has come to take a stand against this nonsense. The reason I am so irrevocably at war against evolution is because it teaches the dangerous nonsense that there is no GREAT SCIENTIST (God). I would have more respect for them if they honestly admitted they simply are not sure.

The interesting scientist and writer Stephen Jay Gould—in his fascinating book, *Wonderful Life* wrote what he believed was the answer to a superb question; "Why do humans exist?" And so, if you wish to ask the question of the ages—why do humans exist?—a major part of the answer, touching those aspects of the issue that science can treat at all, must be: because *Pikaia* survived the Burgess decimation."

(Pikaia is a type of fish/chordate, from the Burgess Shale which is high in the Canadian Rockies.)

In other words, 'Because a two inch fish/worm with a primitive backbone once existed.' In that final paragraph, he also admits; "I do not think that any 'higher' answer can be given, and I cannot imagine that any resolution could be more fascinating."

And there it is—an erudite thinker, cannot think of a better ('Higher!') answer, nor is he, on his own admission able to 'imagine' that any resolution could be more fascinating!

William Blake the 18th century poet—artist and thinker said, "Imagination is evidence of the divine." However, I cannot think or imagine of virtually anything more ludicrously ignoble, than at the conclusion of a fascinating book to suggest our ancient ancestor was a worm/fish. . . . It is not credible, not believable, and is not reasonable.

I can *imagine* a creator delighting in his artistic genius and creating for a purpose. I can *visualise* an earth being prepared through brilliant cosmological constants and silently awaiting the arrival of mankind. I can *think* of God's personality as revealed in nature and possibly grasp some reflected amazing aspects of it. I can *wonder* at what "In the beginning God created the heavens and the earth." might mean in terms of universal time! And not being too concerned as to when that beginning might have been! (I don't actually care.) I can *meditate* on natural and human history and see tenderness and cruelty, hate and love, purpose and chaos, sanity and madness beauty and ugliness, tragedy and triumph, and realise that no woman, or no man has all the answers. I can *ponder* on the past—present, future and realise that unless there is some way of knowing for certain:

There very likely is a God!

It is possible he has communicated at sometime in the past with someone.

That communication is likely to be recorded somehow and somewhere!

I sure would like to know!

How can I find out?

Is it really possible?

Could it be a book?

The one thing I refuse to do is to accept speculative theories as facts. In conclusion I ask my family—friends and readers to very carefully consider, always, what is presented as evidence. When anyone, regardless of educational qualifications speaks of what might have happened on our earth or in our cosmos one million years (or more) ago, *they are guessing!*

The mind is a universe of complexity and the mind of God cannot be analysed, penetrated, understood, probed, examined, caricatured or copied. Enough to have truly lived, loved, laughed, learned, and seen beauty on our earth.

If there is a God, then there surely is hope that one day he shall rectify what obviously went wrong in the past, it is his universe, his earth, his right, his pleasure and ultimately his purpose.

John J. May
March 1st, 2010

"*Follow the evidence wherever it may lead.*"
SOCRATE'S

"*Do to others as you wish them do to you.*"
JESUS

Image Index

The Scream
—Edvard Munch, Norwegian painter. (1863 -1944)   11

The Winged Fantasy
—Great Ages of Man—Ancient Egypt—Time Life,
1972 by Lionel Casson   15

Cat
—Encyclopaedia Britannica, 1973   27

Coca Coca Bottle
—Origin USA   27

Campbells Soup Can
—Origin USA   27

Tree
—Trees—James Underwood Crockett
and Editors of Time Life Books, 1978   27

Bird
—Encyclopaedia Britannica, 1973   27

Fish
—The Ascent of Man, J. Bronowski. BBC   1973   27

Statue/Budda
—Encyclopaedia Britannica, 1973   27

Cat
—Encyclopaedia Britannica, 1973   28

Ape to Man
(The Greatest Deceit in the History of Science)
Icons of Evolution
Science or Myth?
by Johnathan Wells,
Regnery Publishing Inc. USA 2002 — 67

Gorilla
—no connection to—Charles Darwin.
Life on Earth, David Attenborough,
Readers Digest Edition, BBC, 1979 UK. — 84

Earth — 103

Blastocyst.
A Child is Born,
Lennart Nilsson. Dell Publishing,
New York USA 1993. — 110

Baby
(my grandson Logan) Dublin Ireland — 111

Baby with parents
A Child is Born,
Lennart Nilsson. Dell Publishing
NY, USA 1993 — 112

The Winning Sperm
TIME. Great Images of the 20th Century,
Published by Time Books, USA, 1999 — 114

Brain
HUMAN by DK Smithsonian Institute:
Published USA 2004 — 122

Heart
ATLAS of the Human Body:
Published by Chartwell Books Inc. 2008   122

Lungs
ATLAS of the Human Body:
Published by Chartwell Books Inc. 2008   126

Liver
ATLAS of the Human Body:
Published by Chartwell Books Inc. 2008   124

Stomach
ATLAS of the Human Body:
Published by Chartwell Books Inc. 2008   125

Throat
ATLAS of the Human Body:
Published by Chartwell Books Inc. 2008   127

Eye
HUMAN by DK Smithsonian Institute:
Published USA 2004   128

Atomic Bomb
—Symbol of Mankinds Madness   133

Sperm penetrating egg
'A Child is Born'
Lennart Nilsson. Dell Publishing
NY, USA 1993   149

Exact moment of Fusion
'A Child is Born'
Lennart Nilsson. Dell Publishing
NY, USA 1993   149

Embryo
'A Child is Born'
Lennart Nilsson. Dell Publishing
NY, USA 1993   150

Baby in Womb
'A Child is Born'
Lennart Nilsson. Dell Publishing
NY, USA 1993   150

Ramesses I
The USBORNE,
Internet—Linked Encyclopaedia of Ancient Egypt   167

Cell to a Baby   171

Baby
(my grandson Logan)   171

Baby
Genetic Programming that shames
the best computer programme   184

Cell dividing into two
—Lennart Nilsson's 'A Child is Born'
Dell Publishing
NY, USA 1993   186

Blastocyst into a future human being
—Lennart Nilsson's 'A Child is Born'
Dell Publishing
NY, USA 1993   188

Blastocyst
Lennart Nilsson's 'A Child is Born'
Dell Publishing
NY, USA 1993   190

Codes Demystifying Codes
—Lennart Nilsson's 'A Child is Born'
Dell Publishing
NY, USA 1993   195

Baby in womb, our first home
Lennart Nilsson's 'A Child is Born'
Dell Publishing
NY, USA 1993   195

Baby from a single cell
—Lennart Nilsson's 'A Child is Born'
Dell Publishing
NY, USA 1993   196

Baby in water but not drowning
—Lennart Nilsson's 'A Child is Born'
Dell Publishing
NY, USA 1993   200

Peace in the Heart
—Lennart Nilsson's 'A Child is Born'
Dell Publishing
NY, USA 1993   201

Baby, I'm Alive and I've arrived
—Lennart Nilsson's 'A Child is Born'
Dell Publishing
NY, USA 1993   205

The result of love   *208*

The Mysterious Magical cell
—Life, How Did it Get Here?
Published by the Watchtower Society,
NY, USA 1985   210

Chemical Codes; Our lifeline
Vigue-Martin,
Published by Chartwell Books Inc. 2008   211

Earth
—Our one and only home.   213

Human Cells   217

Hair Follice, Skin
ATLAS of the Human Body.
Published by Chartwell Books, 2008   217

Red blood cells—Skull Bones
Vigue-Martin,
Published by Chartwell Books Inc. 2008   218

Heart—Brain
ATLAS of the Human Body
Vigue-Martin, Published by
Chartwell Books Inc. 2008   218

The Greatest pump in History, The Heart
Internet   220

Bibliography

Thomas Paine
—English/American—Author "Rights of Man"
(1737—1809)   v

Henry David Thoreau
—American Writer—Poet—Philosopher
(1817—1862)   ix

Ludwig Wittgenstein,
Austrian Philosopher
(1889–1951)   iii

J.R. Robinson
(Cleric) "Mass & Modernity"   iv

Constantine
—Roman Emperor—
"Called Council of Nicaea" 325AD
(272–337AD)   1 - 2

Charles Dickens
—English Author, "A Christmas Carol"
(1812–1870)   2

Voltaire
—French writer—Philosopher—Playwright
(1694–1778)   22

Sam Harris
—Neurophilosopher Author "The End of Faith"
(1967-present)   5

Cecil Roth & Geoffrey Wigoder
"The Jewish Encyclopaedia"
 pub; 1970   6

Oscar Wilde
Irish Playwright—Poet and Author
"The Picture of Dorian Gray"
(Oct 1854-Nov 1900)   6

Christopher Hitchins
—English/American Polemicist—
Author "God is not Great"
(1949—present)   6

C.T. Russell
–American religious bible restorationist teacher/
started end of the world religion.
(now called Jehovah's Witnesses)
Publisher of Watchtower Magazine
(1852- 1916)   9

Jesus
—Jewish Rabbi and Moral Teacher (5 B.C.E.—30 A.D.)
THE MOST INFLUENTIAL
HUMAN BEING IN HISTORY   9

Maimonides
—(Moses) Spanish Philosopher—Jewish Theologian
(1138–1204)   9

Moses
—Egyptian—Ethical Lawgiver.
Author; Genesis / Exodus/Liviticus/Numbers/ Deuteronomy.
Religious Leader.
Produced the 10 most famous Commandments
of all time.
(1392—1272 B.C.E.)   9

Rick Warren
—American Religious Evangelical Preacher
—Author "The Purpose Driven Life."
(1954-present)   10 - 11

Solomon
—King of the Jews—Sage—Poet—Composer—
Horticulturist—Ornithologist
-Architect–Judge—Dendrologist—Zoologist—Taxonomist—
Husband to 700 wives and 300 concubines.
(10th Century B.C.E.)   17

Gandi—(Mohandas)
Indian; Hindu; Political leader and activist.
PACIFIST.
(1869–1948)   19

Langston Hughes
—American Poet
(1902—1967)   24

Bram Stoker
—Irish Author of Dracula
(1847—1912)   26

Anthony Flew
Ex-life long athiest, who wrote in 2007
the excellent book: 'THERE IS (NO) A GOD'
Published by Harper Collins   29

John Meadows Rodwell
—Christian clergyman—Koran Translator—1861   31

Mohammed
—Founder/Leader of Islam
—Diplomat—Merchant—Philosopher—Orator
—Legislator—Reformer—Military General.
(570—June 8th, 632 A.D.)   31

Adolph Hitler
—Author "Mein Kamph"
Austrian/German political leader of Nazi party.
Preached pathological hatred of Jews and many others
1889—1945)   38

Richard Dawkins
—Kenyan—Author "River Out of Eden."
Atheist—Lecturer(1941-present)   39

Dr. Wernher Von Braun
—German Scientist—
Worked after 1945 for USA Governement.
(1912—1977)   40

Ms Lynn Margulis
—American Professor of Biology. Proctor
(1938- present)   40

Michale J. Behe
—Author; "Darwins Black Box."
Named by National Review and World Magazine
as one of the 100 most important books
of the 20th Century
(1952- present)   40-41-57

Diocletian
—Roman Emperor (284- 305 A.D.)   41

Daniel Lord Smail
—Author; "Deep History and the Brain"
Professor of History, Harvard University, USA
(1961- present)   41

Stephen Jay. Gould
—American Science writer-—Evolutionist—
Palaeontologist—Athiest.
(1941—2002)   41

Simon Sebag Montefiore.
—British historian and Author ;"Young Stalin."
(1965—present)   42

Joeseph Stalin
—Russian Atheist Leader /Dictator
(1878—1953)   42

Charles Darwin
—English Naturalist—Taxonomist
—Author "On Origin of Species"
(1809—1882)   39

John Stuart Mill
—English Author ;
"The Autobiography"
—Philosopher—Political Theorist.
(1806—1873)   43

Cyril Aydon
—Author;
"Charles Darwin—His life and Times."
Published in the UK
by Constable & Robinson Ltd, 2002   44

Christopher Columbus
Italian/Portuguese Explorer—Colonizer—
Discovered America
(1451- 1506)   45

Louis Pasteur
—French Scientist—Chemist—Microbiologist—
Discovered Penicillin
(Disproved the ridiculous theory of
"Spontaneous Generation)
(1822–1895)   45

Sir Isaac Newton
—English Scientist—Mathematician—Physicist
(1642—1727)   45

Thomas Malthus
—British scholar- wrote essay on the
"Priniple of Population."
Huge influence on Charles Darwin
(1766—1842)   46

John Farndon
—Contemporary English Author;
"The Great Scientists."
Published by Arcturus Ltd. 2005   46

Richard Dawkins
—Oxford Professor—Atheist—Writer
(1941—present)   47 - 50

New Scientist—Editorial;
Editor in Chief , Jeremy Webb.
Magazine Editor—Roger Highfield.
On-line Editor—Sumit Paul Choudhury.
Deputy magazine Editor—
Graham Lawton   48

W.R. Thompson
—Director of Commonwealth Institute
of Biological Control UK   48

Francis Hitching
—English Author
"Where Darwin went wrong
and The Neck of the Giraffe"—Dowser.   49

Professor Louis Bounoure
—Biologist   49

Dr. T.N. Tahmisian
—Physiologist—Associate member—
United States Atomic Energy Commission   49

Paul Lemoine
—French Scientist
—Director of Natural History Museum, Paris
(1957)   49

Thomas Huxley
—English Atheist—
"Darwins Bulldog."
(1825—1895)   53

Samuel Wilberforce
—English—"Bishop of Oxford."
(1805—1875)   53

Robert A. Burton
M.D. American Novelist ;
"On Being Certain."
Published 2008

Christopher Hitchins
—English/American Author;
"God Is Not Great."
Atheist
(1949—present)   57

Michael J. Behe
—Biochemist—Biologist—
Professor of Biological Sciences at Lehigh University,
Pennsylvania USA
—Author "Darwins Black Box."
Published by FREE PRESS 1996 New York, USA
Note: this is the most intellectually satisfying book
I have ever read on this topic)   57

Sir J.J. Thompson
—British Physicist
—Nobel Prize in Physics 1906
—Discovererof the Electron
(1856 -1940)   58

Stephen Jay Gould
—American Biologist—Science historian—
Author; wroteEssay "The Panda's Thumb."
and "Ontogeny and Phylogeny"
(1941 -2002)   59

Christopher Booker
—English journalist—"Evolutionary Writer."
London Times
Co-founder of Private Eye Magazine.   59

Stephen W. Hawking
—British mathematician—Theoretical Physicist
—Cosmologist—Philosopher—Scientist—
Thinker—Author; "A Brief History of
Time." and "The Theory of Everything."
(1942—present)   59

William Thorpei
—British Zoologist—Ethnologist—Ornithologist
—Behavioural Biologist
(1902—1986)   60

Fred Hoyle / Chandra Wickramasinghe
—Hoyle was a British Cosmologist—
(1915 -2001)
Both were Astronomers—
Chandra was born in Ceylon in 1939
And was a professor of applied mathematics. 60

Lennart Nilsson,
M D—Sweedish Author/Photographer;
"A Child is Born"
Published by Dell, USA 1990.
Honorary Doctor of Medicine at
The Karolinska Institute, Stockholm.
Co-labourer Mr. Lars Hamberger M D. Professor and
Chairman of Obstetrics and Gynaecology
at Gothenburg University, Sweden 61

Dr. Francis Collins
—MD—PH. D—American Author;
"The Language of God"
Published by FREE PRESS,
2006 New York, USA.
Director of the National Human Genome
Research Institute. (1950—present) 62

Phil Town
—American Author; "Rule Number One" (
Quoting Boswell in,
"The Life of Samuel Johnson.) 62 - 63

Dr. George Wald:
American Professor Emeritus of Biology,
Harvard University
Nobel Prize winner of Biology 1971.
 Biochemist. (
1906—1997) 63

Titus Lucretius
—Roman Epicurean,
Poet/Philosopher
(99—50 B.C.)   64

Stevan Weinberg
—American scientist
in conversation with Richard Dawkins
Shared Nobel prize in physics.
Received extraordinary number of awards.
Born 1933—present   65

Shakespeare
—English Playwright.
Actor. Poet. Writer. Re: "Old age"
(1564—1616)   66

Oscar Wilde
—Irish Poet. Playwright.
Wit. Re: "Growing up"
(1854—1900)   66

Josephine May
—Irish Mother of Author, John J. May.
Re: Old age—"No wrinkles on hearts"
(1917- 1996)   66

Sigmund Freud
—Austrian—Hungarian—Neurologist—
Psychiatrist—Psychoanalyst—
Promoter of cocaine for depression
and as an analgesic.   67

Epicurus
—Greek Philosopher,
(341–270 B.C.E.)   67 - 68

Dr. I. L. Cohen
—Author; "Darwin was Wrong"
Published 1984
"The Greatest Deceit in the History of Science."   68

Dr. Wolfgang Smith
—American Author; "Origins Answer Book."
and "The Quantum Enigma."
Philosopher of Science—
Metaphysician. Mathematician.
(1930—present)   68

Micheal J. Behe
—American Biologist/Chemist/
Author of "Darwins Black Box."
Published 1996 by Simon & Schuster
New York, USA   74

Ptolemy
—Roman Mathematician—Astrologer—
Geographer—Poet
(90—168 A.D.)   75

Erasmus Darwin
—Darwin's Grandfather—
English Author; "Zoonomia."
Philosopher—Physiologist—Abolitionist
—Inventor—Poet—Naturalist—
(1731—1802)   76

Adrian Desmond & James Moore
—Authors "Darwin the life of a Tormented
Evolutionist." Published 1994 (British)   76

243

Charles Lyell
—English Author;
"Principles of Geology."
Lawyer—Geologist—Uniformitarian
(1797—1875) 77

Thomas Malthus
—English Clergyman/Author;
"Essay on the Principle of Population."
6th Edition had major influence on
Charles Darwin & Alfred
Russell Wallace. 77

Janet Browne
—Autobiography of Charles Darwin—
"Voyaging." Professor of History,
Harvard University, USA.
B A in Natural Sciences from
TrinityCollege Dublin, Ireland 79

John Fordyce
—Author; "Aspects of Scepticism." 1883 81

Karl Marx
—Prussian Author of "Das Capital"
Published 1867 in German.
Intellectual founder of Communism
with Frederick Engles.
Historian—Political theorist—Sociologist
-Communist—Revolutionary. (1818—1883) 82

Richard Dawkins
—Born Nairobi, Kenya.
Author "The Blind Watchmaker"
(An Oxymoron)
Radical atheist. (1941—present) 82

Charles Darwin
—English Author; "Descent of Man."
Taxonomist. Nature lover.—Evolutionist
—Scientist. He influenced his half cousin
Francis Galton, who
developed biometrics
and coined the unscientific
discredited term "Eugenics."
(1809—1882) 83

Hermann Rauschning
—German Author; "The Voice of Destruction."
Knewand associated with Adolf Hitler.
(1887—1982) 83

Oscar Wilde
—Irish intellectual.
Author and Playwright,
 "Lord Arthur Savilles
Crime." And "The Importance of Being Ernest."
(1854 -1900) 87

Albert Einstein
—German—Jewish–Scientist—Intellectual.
"In conversation."
Mathematician. Famous for his
 theory of relativity.
(1879—1955) 87

Hans Christian Andersen
—Danish children's writer
"The Emperors New Clothes."
(1805—1875) 87

Sean O'Casey
—Irish Playwright—
"The Plough and the Stars." Declined order
of the British Empire.
(1880- 1964)   95

Sir John Hershal
—British Astronomer in conversaion
Re:"That Mystery of Mysteries."
Mathematician—Chemist—Botonist—
(1792—1871)   100

Louis Pasteur,
French scientist demolished
unscientific notion of
—"Spontaneous Generation."
Microbiologist—Chemist.
(1822—1895)   101

Oscar Wilde
—Brilliant Irish writer
"On Belief. "
(1854 -1900)   104

Andrew Newberg
MD—Psychiatrist & Mark Robert Walkman,
Therapist—Neurologist—
Authors; "Why We Believe What We Believe."
Published by
Free Press New York, USA 2006   104

The Barna Group
—Sociological Study.
Research organisation (U.S.A)   107

Lennard Nilsson
—Sweedish Author "A Child is Born."
PhotographerExtraordinaire.   109

Michael Shermer
—American Author "Why Darwin Matters."
Published by
Time Books. First Owl Books Edition,
2007, USA.
Publisher of SkepticMagazine.   109-115

Lynn Margulis
—American Professor of Biology,
(1938—present)
"Challenge to the worlds molecular biologists."   115-135

Ludwig Wittgenstein
—Austrian Philosopher, On Resurrection.
(1898—1951)   134

Dr. T.N. Tamisian
—United States Atomic Agency Commission
"Evolution and the Emperors New Clothes."   136

Jerry Coyne
—American professor of Biology.
Geneticist. Ecologist "On Bias."
(1949—present)   139

Professor Louis Bounoure
—Swiss Scientist.
Quote from "The Advocate."   144

Malcom Muggeridge
—English Journalist—Author—Philosopher
Re: "Oneof the greatest jokes in human history."
(1903- 1990)   144

Michael J. Behe
—American Biologist.
Author "Darwins Black Box."
Published 1996   145

Sir Fred Hoyle
—British Astonomer;
"On chance."
Rejects the big bang
theory. (1915—2001)   145

Dr. Carl Sagan
—American Astrophysicist
—Author—Astronomer—Science
Writer. Won Pulitzer Prize for non-fiction.
1978. (1934—present)
"On mathematical probability."   146

Dr Emile Borel
—French mathematician
—Politician on "The laws of probability."   146

Michael Denton
—Brittish/Australian Author;
"Evolution—A Theory in Crisis."
Molecular Biologist on the myth
of "Prebiotic Soup."
(1943—present)   146

Charles Darwin
—English Naturalist.
Taxonomist. Author:
On his "Execrable
book." ... 152

Charles Darwin
—On "Our Eyes." ... 152

Dr. Edwin Conklin
American Biologist—
Zoologist. On the facetiousness of
"Chance." (1863—1952) ... 153

Dr. George Wald
American Professor of Biology
at Harvard University and
Nobel Prize winner for
Physiology/Medicine 1967
On "There are only two choices."
From—"Origin, Life and Evolution."
Scientific American.
(1906—1997) ... 153

Paul S. Davies
—British Astronomer
—Physicist—Cosmologist—
Astrobiologist on "The Greatest Puzzle." ... 156

Albert Einstein
—German Physicist—Author—
Philosopher on "Universal Order."
(1879- 1955) ... 156

Christopher Hitchins
—English writer—
Controversialist on,
"The Wisdom of Silence."   156

Michael Rose
—American Evolutionist—
Biologist on the "Complexity of Biology."   160

Richard Feyman
—American Scientist—Humanitarian.
Quoted from John Polkinghornes book
"Quantum Theory."
(1918- 1988)   160

Chris Isham
–English Theoretical Physicist
—Mathematician—
Imperial College
London.
Re: "Mixing of Science Fact
with Science Fiction"   160

Sam Harris
—American non-fiction writer—
"The End of Faith."
Neurophilosopher/Advocate
of scientific scepticism.
(1967—present)   172

Scott Adams
—American Author; "Gods Debris."
Economist—Mensa member   176

Kent Hovind
—American Christian Creationist.
Re; "His famous Offer . . ." Science
teacher.
Currently in prison in USA (2010)
for tax evasion.
(1953—present) 182

Lennart Nilsson
—Swedish Author "A Child is Born."
Published 1990 by Dell.
A DTP/Seymour Lawrence Book;
Translated into English by Clare James. 186

Stephen Jay Gould;
Harvard Palaeontologist
—Evolutionary Theorist—writer
of more than 20 books,
300 Essays and 1000 technical papers.
Published his
1433 page,
"The Structure of Evolutionary Theory."
and said "Error is the inevitable
by-product of daring." 212

Bill Gates
—American co-founder with Paul Allen
of Microsoft.
Philanthropist
—Author.
Re: his famous quote on
"DNA complexity."
(1955—present) 214

Richard Dawkins
—East African Author
"The selfish Gene."
Brilliant orator.
(1941—present)   214

Stephen Jay Gould
American Scientist
—Atheist—Author.
"Wonderful Life."
World Famous Palaeontologist. (1941 -2002)   223

William Blake
—British writer—Thinker—
Enlightenment poet—
18th Century
Artist.
(1757 -1827)   223

P.S
My Father was fond of this facinating quote by a man he admired and respected immensely;

JESUS........"The truth shall set you free."